Math for Real Life

T0197956

Math for Real Life

*Teaching Practical Uses
for Algebra, Geometry
and Trigonometry*

JIM LIBBY

McFarland & Company, Inc., Publishers
Jefferson, North Carolina

Library of Congress Cataloguing-in-Publication Data

Names: Libby, Jim, 1955–
Title: Math for real life : teaching practical uses for algebra,
geometry and trigonometry / Jim Libby.
Description: Jefferson, North Carolina : McFarland & Company, Inc., 2017. |
Includes bibliographical references and index.
Identifiers: LCCN 2016052898 | ISBN 9781476667492
(softcover : acid free paper) ∞
Subjects: LCSH: Mathematics—Study and teaching. |
Algebra—Study and teaching. | Geometry—Study and teaching. |
Trigonometry—Study and teaching.
Classification: LCC QA135.6 .L54 2017 | DDC 510.71—dc23
LC record available at https://lccn.loc.gov/2016052898

British Library cataloguing data are available

ISBN (print) 978-1-4766-6749-2
ISBN (ebook) 978-1-4766-2675-8

Front cover images © 2017 iStock

Printed in the United States of America

McFarland & Company, Inc., Publishers
Box 611, Jefferson, North Carolina 28640
www.mcfarlandpub.com

For Donna

TABLE OF CONTENTS

Preface

Many people react much as the mother on the television show *The Middle* did when her son got a D on a math paper: "Math is very important in life. You use math in everything…. Oh even *I* can't say it like I believe it." What are students thinking when they say, "Where are we ever going to use this?" Yes, it could be a cynical ploy and not really a legitimate question. Maybe it's an attempt to divert the teacher's attention away from the math at hand. Maybe it's just a chance to stir things up so it gets a little interesting.

Yet, even if there are ulterior motives, most students really do wonder what that answer is. For each student who is daring enough to ask it, there are plenty of others who are thinking it. This question doesn't seem to come up in English classes—or in science, or government, or band, or health class. All of us tend to shy away from parts of our lives that are difficult and pointless. There isn't much to be done about the difficulty of mathematics, but mathematics is far from pointless. An argument can be made that mathematics is the most useful development in the history of mankind. Yet for many potential mathematicians sitting in high school classes, ironically, it seems without purpose.

In spite of that, it can be difficult for secondary math teachers to come up with answers to the question "Where are we ever going to use this?" There are several reasons for this difficulty.

- Much of the math that is taught in high school needs additional knowledge to be applied. There are many applications available to the individual that understands calculus, although very few high school students will have had calculus. Or, perhaps a greater knowledge of the area of the application itself is needed. A math teacher seeking to understand the formula for electrical resistance in a series would probably need to spend a lot more time understanding the electronics than the mathematics.

1

- Math teachers might be able to draw on past knowledge of applications to answer the question. Many, though, may not have had applications in their background. Current teachers may have had their own high school math teachers that weren't sure about real life applications. Also, potential teacher graduates may be well versed in mathematics, but have, at best, a smattering of classes in physics, chemistry, or economics—the places where real-life applications actually exist. Applications are found in many different subject areas, but that is part of the problem. A math teacher who knows quite a bit about the theory of relativity might know almost nothing about baseball statistics. One is rarely knowledgeable in all areas.
- Understandably, many teachers hesitate to take time out for extra material. Presenting applications takes a certain amount of time and there are ever increasing demands on teachers to teach the pure mathematics of the course. Few math teachers get to the end of the school year feeling they covered everything they wanted. While this will always be a concern, this book will show how it is possible to present applications with a minimum of extra time being spent.
- Finally, it is just human nature that we forget things. Maybe we know quite a few applications. The problem is that perhaps an application suited to the math worked on in February may only occur to us in April. "Oh, I wish had remembered that then." While you will likely find applications in this book that you were not aware of; if nothing else, perhaps this book will help with jogging your memory.

At the risk of starting negatively, let's look at what are not good examples of applications.

Too difficult. There will be plenty of time for pushing students, but this shouldn't be one of them. To present a real world situation that leaves a portion of the class confused defeats the purpose. Students will come away thinking, "OK, I can see how someone could use this stuff, but clearly I won't be one of them." At least with applications, it is probably best err on the side of too easy.

This looks like an application, but it's not. The following problem was taken from a high school math book.

Applications—A metallurgist has to find the values of θ between $0°$ and $360°$ for which $sec(\theta) + csc(\theta) = 0$. Find the solutions graphically.

In what sense is this an application? Why would a metallurgist do this? Isn't this just another math problem with the word "metallurgist" thrown in? And by the way, what is a metallurgist? Students probably will see through this so-called application. Excuse the cynicism, but one might have the suspicion that the authors of the book know it is a selling point to show how math is used in the real world, so they include some problems like this.

Most story problems. Story problems have their place in math classes. Getting students to take a situation and translate it to a mathematical format and solve it is valuable. Likewise, puzzles and various brain teasers can accomplish that same task. *"Three even integers add to 216. What are they?"* That is a fine problem, but it isn't a real world application. *"A room is to have an area of 200 square feet. The length is three times the width. What is the length and width?"* Would this situation come up in real life? The person with this question obviously had a ruler or something if he knew one side was three times as long as the other. Why doesn't this guy just use his ruler to measure the length and width if he really wants to know?

Math so you can do more math. This isn't going to be a big selling point. *"One use of determinants is in the use of Cramer's rule."* Student: "I've learned this math so I can do more math. Thanks." Technically, it counts as an application and examples like this are in this book, but we're skating on thin ice here.

Too long or too involved. We want the applications to be ones with which a large majority of the class understands and feels comfortable. There are times a real world situation could be worked into a major project. An Algebra I or Geometry student, given distances from the sun, could use proportions to create a scale drawing of the solar system. An Advanced Math student could also find the eccentricities of the orbits of the planets and comets of our solar system. This could be time well spent. However, the more involved a project is, the more the chance a student will get bogged down and miss the big picture. Theorem: The amount of complexity in a project is inversely proportional to the number of students that comprehend it. A two-minute presentation by the teacher could accomplish a goal as well as taking several days out for a project.

So how should a teacher present real world situations to a class? From the student's point of view, the answer to "Where are we going to use this?" can probably be accomplished in one to two minutes. So what are some options?

- A teacher may have to resort to something like, "I know they use imaginary numbers in electronics. I'm not sure exactly how, but I know they do." Not ideal, but better than nothing.
- A couple minutes spent showing the class an application may be all that is needed. Imaginary numbers are used in measuring capacitance and voltage in alternating current. A brief explanation could be given with a problem or two at the board.

- A step beyond this is for the teacher to explain the application, with students working a few examples on their own.
- Tell a story. Those who may be drifting off will often perk up if the teacher launches into a story. Just talking about how math is used can be a great strategy. How did we end up with different temperature scales and who exactly are Celsius, Fahrenheit, and Kelvin? How did they figure out cube roots centuries ago? Even if it is basically a history lesson and the math doesn't exactly apply today, students are often still interested. A teacher doesn't always have to be in front of a class with a piece of chalk or marker in hand working math problems.
- The teacher could choose to develop an application into something more involved, designing an assignment or a multi-day project. There are many equations used to analyze baseball statistics. An algebra class could use the statistics found in the newspaper to evaluate players. Despite the previous caution regarding this kind of activity, it has its place.

The complete answer is perhaps some combination of these. Ultimately teaching is still an art.

Finally, I would like to take this opportunity to thank those that have helped me in the writing of this book. Thank you to the staff of McFarland Publishing for giving me the opportunity to bring my thoughts to you.

Many have read my manuscript, primarily relatives who were supportive of what I was doing and were great proofreaders. Donna, David, Steven and John Libby; Karen and Andrew Howard—thank you for your time, effort, and encouragement.

INTRODUCTION

Make this book suit you. It can be read from cover to cover or used as a resource for when a certain topic is being covered. There are many times when a great application comes to mind, but well after the math topic is in the rear view mirror. This book might give information new to you or simply serve as a resource to fit applications to the appropriate places in the school year.

The primary audience for this book is secondary math teachers (although anyone who has ever asked the question, "Where are we ever going to use this?" might find this book interesting reading). At the risk of insulting your intelligence, answers or steps to work out problems have been supplied throughout this book. I have confidence in your ability to work these out on your own. They are supplied only in the interest of clarity and of saving you some time.

The problems here are not meant to be overly rigorous and take away from the main goal of the book. Technically, there will be times when statements like "neglecting air resistance" or other such qualifications should be used. When it doesn't create a major problem, those types of statements will usually go unmentioned. The goal here is not mathematical rigor, but getting students to feel more connected to the math they are learning.

Typically, units are in the English System. Like it or not, most students still have a better feel for this than for the metric system. There are times, such as certain scientific examples, when the metric system is the best to use. At times the system used is somewhat arbitrarily chosen.

To help in finding material, topics have been grouped into typical secondary classes: Algebra I, Geometry, Algebra II, Advanced Math, and Trigonometry. Obviously, there is some overlap and review in these courses, so topics have been placed generally in the courses in which they most commonly appear. For the most part, I have left out the areas of probability and statistics. Most students can easily identify areas of their application.

Learning mathematics is, at times, a confusing, painful, joyless endeavor for many. Hopefully students will come to feel just a little more energized by seeing real world applications to this hard work they are doing.

I

ALGEBRA I

Negative Numbers

Students might feel that the concept of numbers less than zero just doesn't make sense. They wouldn't be alone. There have been several classifications of numbers, including negative numbers, that were not been well received initially. Negative numbers have existed in various forms for a couple of thousand years, but didn't really catch on in Europe until the 1500s. The Greek mathematician Diophantus (c. AD 250) considered equations that yielded numbers less than zero to be "absurd."[1] So, it is understandable if students today take a while to warm up to them. It is true that numbers less than zero do not make sense in every context, but there are plenty of places where they do.

Numbers less than zero exist in various areas: Temperature (temperatures below zero), golf scores (shots under par), money (being in debt), bank accounts (being overdrawn), elevation (being below sea level), years (AD vs BC), latitudes south of the Equator, the rate of inflation (falling prices), electricity (impedance or voltage can be negative), the stock market (the Dow was down 47 points), and bad *Jeopardy* scores. Launches of rockets run through the number line. T minus 3 means it's 3 seconds before the test. T minus 3, T minus 2, T minus 1, 0, 1, 2... counts the seconds before and after liftoff: –3, –2, –1, 0, 1, 2...

Those real life examples can be used to make sense of some of the rules for computation. The fact that $4 + (-9) = (-5)$ can make sense to students because they know that dropping 9 degrees from a temperature of 4 degrees makes it 5 degrees below zero. Also, if it is currently 17 degrees, but the temperature is dropping 5 degrees a day for the next 7 days, that is the same as saying that $17 + 7(-5) = -18$. That same situation can show that two negatives multiplied together make a positive. Again, suppose this day it is 17 degrees and the temperature is dropping 5 degrees a day. Going back in time, four days ago it must have been $17 + (-4)(-5) = 37$ degrees.

A gallon of antifreeze might be advertised to be effective to −20° F. However, Sven needs to find that temperature in Celsius. This requires the use of the conversion formula and knowing how to compute with negative numbers.

$$C = \frac{5}{9}(F - 32) = \frac{5}{9}(-20 - 32) = \frac{5}{9}(-52) = -28.9° \text{ Celsius}$$

The rules for computation with negatives are sometimes needed when finding mean averages. If the temperature for four consecutive days is −3, 7, −12, and −2, the average temperature is below zero. Finding a football player's yards per carry is found by adding his total yards and dividing by the number of times he has carried the ball. If he had yardages of 2, 3, and a loss of 8, he averages −1.0 yards per carry.

Absolute Value

Students love lessons on absolute value, because it seems so simple. Whatever your answer is, make it positive. Pointless, but simple. Well, it's not that pointless or that simple. The absolute value is important when finding the difference between two values. The leader of a golf tournament is 5 under par. Second place is a distant 3 over par. We can find the lead by subtracting the values. The difference, 3 − (−5), is 8. Or, we could find the difference, −5–3, is −8. The conventional way of expressing the answer is to write it as a positive number.

The distance from numbers x and y can simply be stated $|x - y|$ without concern over which order to subtract to get a positive value. The difference in elevation between Mt. Whitney, 14,494 feet, and nearby Death Valley, 282 feet below sea level, can be found with $|-282-14,494| = 14,776$ feet. This simpler way of finding a difference becomes quite helpful if writing a line of code to instruct a computer program to find differences in values.

Tolerances

No measurement is perfect. How much error is acceptable in a product so it will still function in the way it was intended? An object that is to be manufactured to have a length of 12 meters may turn out to be 11.998 meters long. Is that close enough? Is 11.93 close enough? Companies set tolerance levels to determine the amount of error that would be considered acceptable in a product. Tolerances are often listed in a plus or minus format, although this information could also be written with absolute value.

Examples:

Alumeco is a European company that manufactures a variety of aluminum products. They, like many companies, have set tolerance levels for their products. The company's website contains product information, including tolerances.[2]

1. Among many other items, they manufacture rectangular metal bars. Bars with a width anywhere between 50 and 80 mm have a listed tolerance of ±0.60 mm. In other words, if manufacturing 55 mm bars, bars of length x would be acceptable if the inequality $|x - 55| \leq 0.60$ holds.

2. Any bars from 80 to 120 mm wide have a tolerance of ±0.80 mm. The tolerance of a bar, length y, that is to be 107 mm in width could be written as $|y - 107| \leq 0.80$.

Average Deviation

There are a number of formulas that measure the variability of data. A common one is the standard deviation. However, average deviation is similar and easier to compute. The average deviation simply finds the average distance each number is from the mean.

To find the average deviation, the distance from the mean is found for each piece of data in the set. Those distances are added and then divided by the number of pieces of data. If the mean is 32, we would want 28 and 36 to both be considered positive 4 units away from the mean. Absolute value is used so there are no negative values for those distances.

Example:

A set of data is {21, 28, 31, 34, 46}. The mean average is 32. The average deviation is computed as:

$$\text{Average deviation} = \frac{|32-21| + |32-28| + |32-31| + |32-34| + |32-46|}{5} = 6.4$$

Statistical Margin of Error

As mandated by the U.S. Constitution, every ten years the government is required to take a census counting every person in the United States. It is a huge undertaking and involves months of work. So how are national television ratings, movie box office results, and unemployment rates figured so quickly—often weekly or even daily? Most national statistics are based on collecting data from a sample. Many statistics that are said to be national in scope are actually data taken from a sample of a few thousand. Any statistic that is part of a sample is subject to a margin of error. (In 1998, President Bill Clinton attempted to incorporate sampling in conducting the 2000 census, but this was ruled as unconstitutional.[3])

Example:

On October 3, 2014 the government released its unemployment numbers for the month.[4] Overall unemployment was listed at 5.9 percent. The report also stated that the margin of error was 0.2 percent. Government typically uses a level of confidence of 90 percent. Thus there is a 90 percent chance that the actual unemployment rate for the month was x, where $|x - 5.9| \leq 0.2$.

Distance from a Point to a Line

The distance from a line with equation $Ax + By + C = 0$ to the point (u,v) is:

$$\text{Distance} = \frac{|Au + Bv + C|}{\sqrt{A^2 + B^2}}$$

Example:
The distance from the point $(-2,1)$ to the line $2x - 3y - 4 = 0$ is:

$$\text{Distance} = \frac{|2(-2)-3(1)-4|}{\sqrt{(2)^2 + (-3)^2}} = \frac{|-11|}{\sqrt{13}} \approx 3.05$$

Angle Formed by Two Lines

To find the acute angle, α, of two intersecting lines; substitute their slopes, m and n, into the following formula:

$$\tan(\alpha) = \left| \frac{m-n}{1 + m \cdot n} \right|$$

Example:
The lines $y = x + 5$ and $y = \frac{8}{7}x + 2$ are nearly parallel, having slopes of 1 and $\frac{8}{7}$. Substituting them into the formula gives the following:

$$\tan(\alpha) = \left| \frac{1 - \frac{8}{7}}{1 + 1 \cdot \frac{8}{7}} \right| = \left| \frac{-\frac{1}{7}}{\frac{15}{7}} \right| = \frac{1}{15}$$

Using the arctangent function, α is approximately 3.8 degrees.

Richter Scale Error

The Richter scale is used to measure the intensity of an earthquake. However, like many measurements, there is a margin of error that needs to be considered. Scientists figure that the actual magnitude of an earthquake is likely 0.3 units above or below the reported value.[5] If an earthquake is reported to have a magnitude of x, the difference between that and its actual magnitude, y, can be expressed using absolute value: $|x - y| \leq 0.3$.

Body Temperature

"Normal" body temperature is assumed to be 98.6° F. For any student that has made the case that anything other than 98.6° prevents their attendance at school, there is good news. There is a range surrounding that 98.6 value that is still considered in the normal range and will allow your attendance at school. Your 99.1° temperature is probably just fine. Supposing plus or minus one degree is safe, an expression could be written $|x - 98.6| \leq 1.0$, which would represent the safe range. Why is the absolute value a necessary part of this inequality? Without it, a temperature of 50 degrees would be considered within the normal range, since $50 - 98.6 = -48.6$, which is, in fact, well less than 1.0.

Functions and Relations

There are countless examples of functions and relations in the real world. Many can be found in other sections of this book. As the following examples show, the domain of a function can consist of one to many variables.

Water Pressure

For every mile descended into the ocean, the water pressure is approximately 1.15 tons per square inch.[6] This could be expressed in function notation as $f(x) = 1.15x$, where x is the miles below sea level.

Examples:
1. The deepest point in the ocean is the Mariana Trench in the Pacific Ocean, near Japan. What is the water pressure at its deepest point, 6.85 miles below sea level? [Answer: 7.88 tons/sq. inch]
2. The *Titanic* lies at the bottom of the ocean, 2.36 miles below the surface. What is the pressure at that point? [Answer: 2.71 tons/sq. inch]

For perspective on these numbers, the bad guy in *Rocky IV* (granted, a fictional character) claimed he could punch with the equivalence of 1.075 tons per square inch.

Four Function Calculators

Basic calculators are often referred to as four function calculators. Addition, subtraction, multiplication, and division are indeed examples of functions.

Addition:	$a(x,y) = x + y$
Subtraction:	$s(x,y) = x - y$
Multiplication:	$m(x,y) = x \cdot y$

Since dividing by zero is not allowed, the domain must be adjusted to for division to be a function.

Division:	$d(x,y) = x \div y$, for $y \neq 0$

Students could be asked to explain why these do qualify as functions and, also, whether they are one-to-one functions.

Carbohydrates, Protein, Fat

When studying the nutritional value of certain foods, the percent of the calories that are obtained from carbohydrates, protein, and fat is important information. Health professionals have stated that a maximum of 20 percent to 35 percent of a person's caloric intake should come from calories of fat.[7]

The label of a recently purchased jar of peanut butter listed the following amounts per serving: Total fat: 12 grams; total carbohydrates: 15 grams; and total protein: 7 grams. The ratio of the amount of fat compared to the total would seem to be 12 out of 34, which is 35 percent. Not too bad. This falls into that 20 to 35 percent range. That seems odd, though, because conventional wisdom holds that peanut butter is high in fat. The ratio in question, however, is not that of grams, but of calories. It turns out that each gram of fat (f) is worth 9 kilocalories. Each gram of protein (p) and of carbohydrates (c) is worth 4.[8]

Thus, a function that determines the number of kilocalories from those sources would be:

$$\text{Number of kilocalories} = K(f,p,c) = 9f + 4p + 4c$$

Using this formula, that serving of peanut butter actually contains $K(12,15,7)$ = 196 kilocalories. And since those 12 grams of fat are worth 108 kcal, our fat percentage is up to 108 out of 196 or 55 percent.

Sabermetrics

As the movie *Moneyball* demonstrated, there are many formulas that can be used to judge a baseball player's worth to a team. Some are relatively simple and some are pretty involved, containing several variables. One example is Equivalent Average, which judges a hitter's worth to a team by combining a number of a batter's statistics.

$$\text{Equivalent Average} = \frac{H + TB + 1.5(W + HBP) + SB}{AB + W + HBP + CS + \frac{SB}{3}},$$

where H = Hits, TB = Total Bases, W = Walks, HPB = Hit by pitch, SB = Stolen bases, AB = At bats, and CS = Caught stealing.

Exercise Heart Rates

The number of times a person's heart beats in a minute is a good indication of how hard he or she is exercising. The Karvonen Formula[9] enables exercisers to find the level of exertion that is optimal for them—not too easy and not too difficult. This level of exercise is known as the target heart rate

(THR). The formula is THR = ((MHR − RHR) · I) + RHR. MHR is the athlete's maximum heart rate per minute. A person's MHR is a function of age (A) and is often found with the formula MHR = 220 − A. (However, the Journal of Medicine and Science in Sports and Exercise claims that MHR = 206.9−0.67A is a more accurate formula.[10]) RHR is the athlete's resting heart rate. This is found by simply counting the beats per minute when at rest. I is intensity, expressed as a percentage, and is a somewhat subjective judgment by the individual. Beginners are recommended to have their intensity be in the 50 to 60 percent range.

Example:
A 20-year-old female decides to take up jogging. Thus her intensity should be in the 50 to 60 percent range. Her resting heart rate is 70 beats per minute. In what range would her target heart rate lie?

The maximum heart rate should lie between (200−70)0.5 + 70 and (200−70)0.6 + 70. So, she should aim for a heart rate somewhere between 135 and 148 beats per minute while exercising.

Quarterback Ratings

A number of factors could be used to determine the effectiveness of a quarterback. The NCAA and NFL use a combination of yards (y), touchdowns (t), interceptions (i), and completions (c). These are compared to the number of pass attempts (a), to come up with a single number. The following formula was adopted by the NCAA in 1979.[11]

$$\text{Quarterback Rating} = \frac{8.4y + 330t - 200i + 100c}{a}$$

Example:
The winning quarterback in the 2014 championship game, Jameis Winston, completed 20 of 35 passes for 237 yards. He passed for 2 touchdowns with no interceptions. What was his rating?

$$\text{Rating} = \frac{8.4 \cdot 237 + 330 \cdot 2 - 200 \cdot 0 + 100 \cdot 20}{35} = 132.88$$

The NFL formula is more complex. Also, there are caps on some of the categories. It's a handful. The formula is[12]:

$$\text{NFL Rating} = \frac{5\left(\frac{c}{a} - 0.3\right) + 0.25\left(\frac{y}{a} - 3\right) + 20\left(\frac{t}{a}\right) + \left(2.375 - 25\left(\frac{i}{a}\right)\right)}{0.06}$$

Example:
In the 2016 Super Bowl, Peyton Manning was 13 for 23, for 141 yards. He had no touchdowns, but one interception. What was Peyton's rating? [Answer: 56.6]

There are other ways to rate quarterbacks and things can get ridiculously complex. ESPN has come up with a rating formula that also takes into account a quarterback's running ability and by how many yards incomplete passes miss their intended target.

Basketball Points

The total points scored by a basketball player or team is a function of three variables. A function could be written based on the number of free throws (x), 2-point field goals (y), and 3-point field goals (z) made in a basketball game. It could be written as: $P(x,y,z) = x + 2y + 3z$.

A number of questions could be asked regarding this function. Students could state why this is a function and whether it is a one-to-one function. Students could be asked to find all possibilities for $P(x,y,z)$ to equal a specific value, say 27 points. Or, students could find y, such that $P(7,3,5) = P(6,y,4)$.

Piecewise Defined Functions

Piecewise functions are combinations of functions. The portion of the function that is applied depends on which interval of the domain is being considered. This concept ties to a number of real life situations.

Postage Rates

The postage for first-class mail depends on the weight of the letter. The U.S. Postal Service has charts that state the amount of postage needed for letters of various weights. The following information is for letters weighing up to 3.5 ounces. Letters that weigh not more than 1 ounce costs $0.49 in postage, weight not over 2 ounces costs $0.70, weight not over 3 ounces costs $0.91, and weight up to 3.5 ounces costs $1.12. This information could be written as a piecewise defined function:

$$f(x) = \begin{cases} 0.49, & \text{if } x \leq 1 \\ 0.70, & \text{if } 1 < x \leq 2 \\ 0.91, & \text{if } 2 < x \leq 3 \\ 1.12, & \text{if } 3 < x \leq 3.5 \end{cases}$$

Children's Dosages

Dosage amounts of medicine may be too much for those with a low body weight. Chewable 160 milligram tablets of acetaminophen are available for children that weigh at least 24 pounds. The following table gives information on how many tablets can be given to a child based upon his or her weight in pounds[13]:

Child's Weight	24 to 35	36 to 47	48 to 59	60 to 71	72 to 95	96 and above
# Tablets	1	1.5	2	2.5	3	4

The information contained in this table can also written in piecewise function notation, where x is the child's weight in pounds and g(x) is the recommended number of 160 milligram tablets to be given.

$$g(x) = \begin{cases} 1, \text{if } 24 \leq x \leq 35 \\ 1.5, \text{if } 36 \leq x \leq 47 \\ 2, \text{if } 48 \leq x \leq 59 \\ 2.5, \text{if } 60 \leq x \leq 71 \\ 3, \text{if } 72 \leq x \leq 95 \\ 4, \text{if } x \geq 96 \end{cases}$$

Football Yard Lines

Most students know how the yard lines are marked on a football field. Each end zone is 10 yards long. At the start of the end zone is a goal line, which can be thought of as the zero yard line. From there yard lines are marked 10, 20, 30, 40, 50, 40, 30, 20, 10.

When first learning about piecewise defined functions, an interesting exercise is for students to try the following:

You are standing on a goal line. Write a piecewise defined function that will determine what yard line you are on after walking x yards down the field.

Students might come up with something like this:

$$f(x) = \begin{cases} x, & \text{if } 0 < x < 50 \\ 50, & \text{if } x = 50 \\ 100 - x, & \text{if } 50 < x \leq 100 \end{cases}$$

There are some variations that would work just as well. Students could determine which variations work and which do not.

Students could make predictions as to what the graph might look like and then sketch the graph. Additionally, this exercise could be done again, but now starting at the end line, which is 10 yards from the goal line.

Hurricane Scale

The Saffir-Simpson Hurricane Scale is a 5-point rating scale used to categorize the severity of hurricanes. It was developed in 1969 by an engineer, Herbert Saffir, and a meteorologist, Robert Simpson.[14]

Category	Wind Speed	Effect
1	74–95 mph	Minimal damage
2	96–110 mph	Moderate damage
3	111–129 mph	Extensive damage
4	130–156 mph	Extreme damage
5	over 156 mph	Catastrophic damage

The table gives an example of a relationship that could be written as a piece-wise defined function. For example, one of the pieces of the function could be written f(x) = 1, if 74 ≤ x ≤ 95.

Television Screens and Seating Distances

The Toshiba Company has made recommendations on what size of television screen to purchase based on the distance a viewer would be seated from the screen.[15]

Screen Size	Minimum Viewing Distance	Maximum Viewing Distance
40 inches	4.0 feet	6.3 feet
42 inches	4.2 feet	6.7 feet
46 inches	4.6 feet	7.3 feet
47 inches	4.7 feet	7.4 feet
50 inches	5.0 feet	7.9 feet
55 inches	5.5 feet	8.7 feet
65 inches	6.5 feet	10.3 feet

This could lead to a number of questions or activities. Is this a relation? Is this a function? List ten ordered pairs from the data. Write as a piecewise relation. Graph the relation. Based on the chart, what might be a good range of viewing for a 32-inch television?

Grading Scales

Students are probably familiar with the way their grades are determined. Typically, grades are based on the percentage of points students receive out of the total number of points possible. Whether they realize it or not, the grading scale is a piecewise defined function in which the domain is a percentage and the range is a letter grade: A, B, C, D, or F. Many schools' typical grading scale would look like this:

$$g(x) = \begin{cases} A, \text{ if } 90\% \leq x \leq 100\% \\ B, \text{ if } 80\% \leq x < 90\% \\ C, \text{ if } 70\% \leq x < 80\% \\ D, \text{ if } 60\% \leq x < 70\% \\ F, \text{ if } x < 60\% \end{cases}$$

Students could try to write this function on their own. Students could also make adjustments to reflect changes to this scale. Most students would consider a 79.6 percent to be worthy of a B, since it rounds to 80 percent. If they want that to be the case, how could the function be rewritten? Also, perhaps students could revise their own scale so it includes grades of A+, A-, B+, and so forth.

Tax Brackets

When people say they are in a certain tax bracket, they are actually talking about being on a certain line of a piecewise defined function. The following is a part of the federal 2013 tax table filing as a single individual.
Taxable income is:

At least	But less than	Amount of tax
$30,000	$30,050	$4,058
$30,050	$30,100	$4,065
$30,100	$30,150	$4,073
$30,150	$30,200	$4,080
$30,200	$30,250	$4,088

Students could take the above table and write it in function notation. (e.g., $f(x) = 4,058$, if $30,000 \le x < 30,050$). However, this table only applies to those that have taxable incomes of less than $100,000. The following is used for those with higher taxable incomes.

Again, taken from the instruction book for the Federal 2013 year:

Taxable Income If line 43 is—	(a) Enter amount from line 43	(b) Multiplication amount	(c) Mult (a) by (b)	(d) Subtraction amount	Tax— Subtract (d) from (c)
At least $100,000, but not over $183,250		28%		$6,706.75	
At least $183,250, but not over $398,350		33%		$15,869.25	
At least $398,350, but not over $400,000		35%		$23,836.25	
Over $400,000		39.6%		$42,236.25	

Students could also transform this table into function notation (e.g., $f(x) = 0.28x - 6,706.75$ if $100,000 \le x \le 183,250$).

That function could lead to interesting questions regarding the domain. How much money did you make if your tax was $50,000? Students might decide to solve the equation, $50,000 = .28x - 6,706.75$. This gives a solution of 202,524.11. But, hold on. That equation only applies if x is such that it lies between 100,000 and 183,250. So, although it is a solution to that equation, it isn't a part of the domain of that piece of the function. Students would have to look at other intervals to find the correct answer.

Overtime Pay

Employees often get paid "time and a half" for working more than forty hours in a week. Equivalently, this phrase could be considered "rate and a half." The rate of pay the employee receives changes after working beyond the forty-hour mark. Suppose an employee gets paid $12 an hour. A function that shows his or her weekly pay for x hours of work is:

$$f(x) = \begin{cases} 12x, & \text{if } x \le 40 \\ 12 \cdot 40 + 12 \cdot 1.5(x - 40), & \text{if } x > 40 \end{cases}$$

Add, Subtract, Multiply and Divide Polynomials

Comparing Costs

If f(x) is a function that gives the annual costs of stock A and g(x) gives the annual costs for stock B, those costs can be compared by combing those functions in various ways.

Find a function that describes how much more the cost of stock A was over stock B? [Answer: f(x) – g(x)]

What the total cost of the stocks each year? [Answer: f(x) + g(x)]

Next year, if the cost of stock A is 3 percent higher and stock B is 2 percent higher, what will be the cost of the stocks? [Answer: 1.03 · f(x) + 1.02 · g(x)]

What is the average cost of the stocks each year? [Answer: $\frac{f(x) + g(x)}{2}$]

What is the percentage ratio of Stock A to Stock B? [Answer: $100 \cdot \frac{f(x)}{g(x)}$]

Amount of Daylight

The trigonometric function f(x) could represent the time of sunset on a given day of the year (x) at a particular location. The function g(x) could be the time of sunrise at that same location. A new function showing the amount of the daylight at any time of the year would be represented by the new function f(x) – g(x).

Surface Area to Volume

The ratio of surface area to volume is an important concept. Houses with a high surface area to volume would generally be more expensive to heat and cool. Products with a high surface area to volume require more packaging. Even at the cellular level, the rate of chemical reactions can depend on the surface area to volume ratio of the cell. Finding this ratio can often involve dividing two polynomials and simplifying the result.

Examples:

Solid	Surface Area	Volume	Surface Area ÷ Volume
Cube	$6s^2$	s^3	$\frac{6s^2}{s^3} = \frac{6}{s}$
Rectangular Solid	$2lw + 2lh + 2wh$	lwh	$\frac{2lw + 2lh + 2wh}{lwh} = \frac{2}{h} + \frac{2}{w} + \frac{2}{l}$
Right Cylinder	$2\pi rh + 2\pi r^2$	$\pi r^2 h$	$\frac{2\pi rh + 2\pi r^2}{\pi r^2 h} = \frac{2}{r} + \frac{2}{h}$
Right Cone	$\pi rs + \pi r^2$	$\frac{1}{3}\pi r^2 h$	$\frac{\pi rs + \pi r^2}{\frac{1}{3}\pi r^2 h} = \frac{3s}{rh} + \frac{3}{h}$
Sphere	$4\pi r^2$	$\frac{4}{3}\pi r^3$	$\frac{4\pi r^2}{\frac{4}{3}\pi r^3} = \frac{3}{r}$
Torus	$\pi^2(b^2 - a^2)$	$\frac{1}{4}\pi^2(a + b)(b - a)^2$	$\frac{\pi^2(b^2-a^2)}{\frac{1}{4}\pi^2(a+b)(b-a)^2} = \frac{4}{b-a}$

Profit Function

A business collects a certain amount of money from sales. This business must pay its expenses out of this. Everything left is profit. The profit function is found by subtracting the cost from the revenue function. In economic theory, this concept can be expressed as the equation $P(x) = R(x) - C(x)$.

Example:

These functions would likely be more complex than this, but here is an example. A business is selling an item for $50 each. So, $R(x) = 50x$. The company has fixed costs (rent, utilities, etc.) of $2,000 per month and the cost to produce each item is $20. So, $C(x) = 20x + 2,000$. The profit function for the month would be:

$$P(x) = R(x) - C(x) = 50x - (20x + 2,000) = 30x - 2,000$$

Also, dividing these functions by the number produced finds the average profit, cost, and revenue. For example, the average profit for each item would be: $\frac{P(x)}{x} = \frac{30x-2,000}{x} = 30 - \frac{2,000}{x}$.

Musical Harmonics

Trigonometric functions are probably beyond the scope of most Algebra

I students. However, the concept of musical overtones can be presented in a general way.

A plucked string can vibrate along its full length and, at the same time, vibrate in two, three, and more sections. This can be described by the combination of a number of trig equations. The first harmonic, or fundamental tone, might be $f(x) = 20\sin(220x)$. The following harmonics could be $g(x) = 16\sin(440x)$, $h(x) = 13\sin(660x)$, and $j(x) = 9\sin(880x)$. The complete function representing the sound produced could be written as $t(x) = f(x) + g(x) + h(x) + j(x)$.

Composition of Functions

Inverses

One of the most important concepts in mathematics is that of the inverse function. Two functions, f and g, are inverses if what f does, g undoes. To find if two functions, f and g, are inverses, one must show that $(f \circ g)(x) = x$ and $(g \circ f)(x) = x$.

Geometric Transformations

Geometric transformations are done by companies such as Pixar in the making of animated movies.[16] In years past, each cel was drawn by hand. Now, more likely, images are made up of points located by ordered pairs or ordered triples. These images are then manipulated to show movement.

Transformations of geometric figures can take place singly or one followed by another. Similar to the algebraic case, a composition of two transformations matches a shape to a second shape, which is then matched to a third. A translation of two units to the right and three up, followed by a reflection across the y-axis could be written as: $R_{\text{y-axis}} \circ T_{2,3}$.

Inflation

If prices rise 5 percent annually, the function $f(x) = 1.05x$ shows the relationship of prices from one year to the next. Similarly, if the next year inflation is at 6 percent, that could be expressed as $g(x) = 1.06x$. The two-year change in prices could be found by composing the two functions: $g(f(x)) = 1.06(1.05x) = 1.113x$. An item that cost \$80 after those two years pass, should cost $f(g(80)) = 1.113(80) = \$89.04$.

Discounts

Sometimes stores will have an item on sale and state that this deal is not available with any other discounts. However, sometimes an individual can take advantage of multiple discounts. This person would, in essence, be composing functions.

Example:

You have two coupons that could be used on an item. One is for $20 off. Another is for 10 percent off. They could be applied in either order. How could a function be written to express these situations, and which would be the best way to apply the coupons?

Since 10 percent off is the same as paying 90 percent of the cost, the 10 percent off coupon could be expressed as f(x) = 0.9x. The $20 off coupon could be written as g(x) = x – 20. Using the 10 percent off coupon first, then taking $20 off could be expressed as g(f(x)) = g(0.9x) = 0.9x – 20.

Using the $20 off coupon first could be expressed as f(g(x)) = f(x – 20) = 0.9(x – 20) = 0.9x – 18.

Comparing the two results shows that applying the 10 percent off coupon first will result in a larger discount.

Currency Exchanges

Traveling between countries often means having to exchange money in order to use a country's currency. These exchange rates can change from day to day. On November 20, 2014, the following exchange rates applied: Each U.S. Dollar was worth 0.6392 British Pounds. Each British Pound was worth 2.4406 Bosnian Marks. Each Bosnian Mark was worth 0.7262 Canadian Dollars. Each Argentine Peso is worth 1.330 Canadian Dollars. These exchanges could be expressed as functions:

U.S. dollars → British pounds:	f(x) = 0.6392x
British pounds → Bosnian marks:	g(x) = 2.4406x
Bosnian marks → Canadian dollars:	h(x) = 0.7262x
Argentine pesos → Canadian dollars:	j(x) = 0.1330x

The concept of composition of functions can be used to show multiple exchanges.

U.S. dollars → Bosnian marks:	g(f(x)) = 2.4406(0.6392x) = 1.560x
British pounds → Canadian dollars:	h(g(x)) = 0.7262(2.4406x) = 1.7724x
U.S. dollars → Canadian dollars:	h(g(f(x))) = 0.7262(2.4406(0.6392x)) = 1.1329x
Bosnian marks → Argentine pesos:	j^{-1}(h(x)) = 7.5188(0.7262x) = 5.4602x

Temperature Scales

There have been a number of different temperature scales. The main ones in use today are the Fahrenheit, Celsius, and Kelvin scales. Olaus Roemer (pronounced "Oh-LAS RO-mer"), a Danish scientist, developed an alcohol-based thermometer in which 7.5 corresponded to the freezing point of water and 60 to its boiling point. Daniel Fahrenheit, a German scientist, began with

Roemer's work and built the first truly modern thermometer in 1714. He used the more accurate substance mercury, rather than alcohol, and expanded Roemer's scale by basically multiplying his numbers by four. After some adjustments, the familiar 32 and 212 were set as the freezing and boiling points, respectively, for water. In 1742, Anders Celsius, a Swedish astronomer, decided a better scale would be to set the freezing and boiling points to be 0 and 100, respectively.[17]

Temperature is a measure of the kinetic energy of atoms and molecules in a substance. The faster those atoms and molecules move, the higher the temperature. That means there could theoretically be a point at which molecules do not move at all. Since you can't go any slower than motionless, there is a lowest possible temperature. That temperature is –273.15°C. In 1848, William Thompson (a.k.a. Lord Kelvin), British inventor and scientist, developed a new scale in which he simply subtracted 273.15 from each of Celsius' numbers. His scale began at what is known as "absolute zero."[18] His formula was K = C – 273.15.

These three temperature conversion formulas could be written in function notation. Those functions could also then be composed to find further conversion formulas.

Given the functions $C(x) = x - 273.15$ (Kelvin to Celsius) and $F(x) = \frac{9}{5}x + 32$ (Celsius to Fahrenheit), a formula to convert Kelvin to Fahrenheit can be developed:

$$F(C(x)) = \frac{9}{5}(x - 273.15) + 32$$

$$= \frac{9}{5}x - 459.67$$

Also, given the formulas $C(x) = \frac{5}{9}(x - 32)$ (Fahrenheit to Celsius) and $K(x) = x + 273.15$ (Celsius to Kelvin) a formula to convert Fahrenheit to Kelvin can be found:

$$K(C(x)) = \frac{5}{9}(x - 32) + 273.15$$

$$= \frac{5}{9}x + 255.37$$

Coordinates

It seems odd that analytic geometry was so long in developing. In various forms, both algebra and geometry had been around a very long time before becoming integrated. Analytic geometry was attributed primarily to Rene Descartes in 1637, and mathematics was never the same after. The Latinized form of Descartes' name is Cartesius, thus giving the name to the Cartesian coordinate plane.

At an early age, students encounter simple bar and line graphs in newspaper and magazines. Data points on those graphs could be considered as ordered pairs and graphed on a coordinate plane. It is helpful for students to see that the x,y-coordinate plane also exists in some other areas they might not have considered.

Screen Resolution

Computer images are made up of pixels. The location of a pixel is determined by its location on a horizontal and a vertical axis. For a screen resolution of 800 by 600, there are 800 pixels on the horizontal axis and 600 on the vertical. The location of a specific pixel can be determined in much the same way it is on the Cartesian plane. The (0,0) point is located in the upper left corner. The 800x600 resolution has pixel locations horizontally from 0 to 799 and vertically from 0 to 599.[19] A pixel's location can then be assigned a particular color depending on how many bits of memory can be assigned to each pixel. If each receives 1 bit of memory, there are 2^1 colors available. Each location can be either black or white. For an 8-bit display system there are 8 bits available for each pixel. So, in an 8-bit display system there are 2^8 = 256 different possible colors for each pixel. In a 24-bit system there are 2^{24}, or approximately 17 million different colors available for each pixel location.[20]

Latitude and Longitude

Map making and the concepts of latitude and longitude predate analytic geometry. Each, however, use the concept of perpendicular axes to locate points. Maps of the world use the equator as its x-axis and the prime meridian as its y-axis. The equator is a naturally occurring location, being contained in the plane perpendicular to the Earth's axis of rotation. The prime meridian was established in 1888 in a rare show of cooperation in which nations of the world agreed to set it at the line of longitude that goes through Greenwich, England.

Motion Capture Suits

Motion capture suits have reflective markers attached in key locations on the suits. Sites such as elbows, knees, and feet are typical locations for these markers. As the individual wearing the suit moves, the coordinates of those markers are recorded on a computer. This technique has been used for special effect work such as with the character Gollum in *The Lord of the Rings* series.[21] It also has applications in areas such as video games, robotics, and analyzing the movements of athletes.

Slope

Mathematically, slope is defined by the ratio, $\frac{\text{Change in y}}{\text{Change in x}}$. Once x and y are defined, the slope concept gives the rate of change and can be tied to many real world applications. If a graph shows the total number of movie tickets sold over a period of time, $\frac{\text{Change in tickets sold}}{\text{Change in time}}$, shows the rate that ticket sales have increased or decreased over that period. There are many rates of change that are familiar concepts. Speed = $\frac{\text{Change in distance}}{\text{Change in time}}$, acceleration = $\frac{\text{Change in speed}}{\text{Change in time}}$, inflation = $\frac{\text{Change in price}}{\text{Change in time}}$, gas mileage = $\frac{\text{Change in distance covered}}{\text{Change in fuel used}}$, impulse = $\frac{\text{Change in momentum}}{\text{Change in time}}$, and depreciation = $\frac{\text{Change in value}}{\text{Change in time}}$. Slope is also the underlying concept of differential calculus. While algebra students may not be ready for this entire topic, it can be shown how slope ties to areas in calculus, such as the concepts of velocity and instantaneous velocity. Following are some other examples of where the concepts of slope and rate of change appear.

Slopes of Roads

At the top of a mountainous road, there is a sign stating the road has an upcoming downgrade of 7 percent. In the language of algebra, this could be stated as having a slope of $-\frac{7}{100}$.

Roads are also built sloping to the middle to allow rain to flow off the road. Regarding the construction of gravel or dirt roads, an Environmental Protection Agency document states that, "Recommendations from supervisors and skilled operators across the country indicate that at least one-half inch of crown per foot (approximately 4 percent) on the cross slope is ideal."[22]

This would correspond to a slope of $-\frac{1}{25}$.

Wheelchair Ramps

In 1990, President George H.W. Bush signed the Americans with Disabilities Act (ADA). Section 4.8.2 of the act states that, "the least possible slope shall be used for any ramp. The maximum slope of a ramp in new construction shall be 1:12. The maximum rise for any run shall be 30 inches."[23]

Examples:
1. If the maximum rise is 30 inches, what is the maximum run?

In order to be in compliance, a ramp with a rise of 30 inches must satisfy the proportion:

$\frac{1}{12} = \frac{30}{x}$ [Answer: x = 360 inches or 30 feet]

2. A ramp of 13.5 feet is planned to reach a patio. The distance from a sidewalk to the patio is 10.5 feet. The patio is 19 inches above the ground. Would a ramp of those dimensions fit the ADA requirements?

After first converting to inches, it is seen the ratio of $\frac{19}{162}$ (= 0.117) greater than the required $\frac{1}{12}$ (= 0.0833) and would not meet the ADA requirement. Some adjustment would have to be made. The run would have to be increased, or the difference between the ground and the patio would have to be changed. This information could be found by solving one of the following proportions. The proportion $\frac{1}{12} = \frac{19}{x}$ could be solved to find the minimum run allowed, or $\frac{1}{12} = \frac{x}{162}$ solved to find the maximum rise allowed.

There are a number of other examples of slope contained within the ADA requirements that can be used as examples; e.g., "The cross slope of ramp surfaces shall be no greater than 1:50."[24]

River Gradient

A river gradient is a way to measure the slope of a river. The gradient has a great deal to do with the river's speed and energy. This has important ramifications for erosion, flood control, and fish populations.

Examples:

1. The Missouri River starts at an elevation of 9,101 feet and, 2,341 miles later, empties into the Mississippi at an elevation of 404 feet.

River gradient is $\frac{9,101 - 404}{2,341}$ = 3.72 feet per mile

2. The Amazon River starts high up in the Andes at 16,962 feet and ends 4,000 miles later at the Pacific Ocean.

River gradient is $\frac{16,962 - 0}{4,000}$ = 4.24 feet per mile

These can be compared to one of the steepest rivers in the world, the Rio Santo Domingo in Guatemala, which descends at roughly 1,900 feet per mile.

Ski Slopes

Students could use the following data to compare various ski slopes. The slope concept could be used to compare the relative steepness of the slopes. Then, the arctangent function could be used to find the angle of depression. Additionally, the Pythagorean Theorem could be used to find the approximate length of the ski run. These are some popular ski slopes in the U.S.

Ski Area	Trail Name	Length (ft)	Vertical Drop (ft)
Taos, NM	Al's Run	2841	1481
Killington, VT	Outer Limits	2241.5	1105
Copper, CO	Sawtooth	1535	588
Squaw Valley, CA	KT-22	1367	588

Topographic Maps

Topographic maps, as opposed to standard maps, give a three-dimensional look to an area. Contour lines are used to show a series of points that have the same elevation and can give the user the sense of rise and fall in elevation. Where lines are closer together, it shows a terrain that is steeper than a region where the lines are farther apart.

Example:

The scale of a map states that 1 inch = 0.2 miles. Also, the map states that each contour line is a change of 40 feet in elevation. Suppose that a trail shown on the map is 1.5 inches in length and that it crosses five contour lines. What is the percent change in elevation?

This problem calls for finding the slope or the rise:run ratio. Since 5 contour lines have been crossed, there is a rise of $5 \cdot 40 = 200$ feet. The run is 1.5 inches or 0.3 miles. The 0.3 mile distance is equivalent to 1,584 feet. So, the rise:run ratio is $\frac{200}{1,584}$, which is a 12.6 percent incline.

Solar Panels

Solar panels collect the Sun's energy and turn it into electricity. To optimize the process, the panels should be perpendicular to the Sun's rays. It so happens that the optimal angle is roughly the same as the latitude where the panels are located. The panels could also be adjusted for the season of the year. The best angle in the summer would be that optimal angle plus 15 degrees. In the winter it would be 15 degrees less than the angle.[25] The amount of slant could be written as degrees of elevation or in slope form. While either is correct, installing the panels on a roof would probably be easier in the slope form. By using the slope ratio, a ruler could be used to measure the rise and run.

Snow Load

Many parts of the United States can get so much snow that a roof collapse is a possibility. Snow load (measured in pounds per square inch) is an important factor in whether a roof will hold up. A person could do the math on his or her own, although it is a bit cumbersome. There are a number of snow load calculators on the internet in which the user substitutes several pieces of information and the snow load is then computed. One of those pieces is the pitch (slope) of the roof. Some call for the pitch to be entered as an angle. Usually, though, it is written as a fraction with a denominator of 12 inches.

Writing Linear Equations

Temperature Formulas

As we've seen previously, in 1714, Daniel Fahrenheit developed a scale for measuring temperature. Although Fahrenheit didn't initially define his scale this way, eventually 32 degrees came to correspond to the freezing point of water and 212 to its boiling point. In 1742, Anders Celsius developed his own scale. He developed a scale in which 100 corresponded to water's freezing point and 0 to its boiling point. (Fortunately, after his death, it was decided to reverse those numbers). The well-known conversion formulas could easily be derived using algebra. Forming a relation of Fahrenheit (x-values) with Celsius (y-values) gives the points (32,0) and (212,100). The slope of the line joining those points is $\frac{100-0}{212-32} = \frac{5}{9}$. Using the point-slope form yields $y - 0 = \frac{5}{9}(x - 32)$ or the more familiar $C = \frac{5}{9}(F - 32)$.

Alternately, the equation $F = \frac{9}{5}C + 32$ could be derived by switching the domain and range, using the points (0,32) and (100,212).

Water Pressure

Pressure from the air at sea level is 14.7 pounds per square inch (psi). Every 33 feet an object goes underwater submits it to another 14.7 psi.[26] This is a linear relationship. An equation for that relationship can be found with this information. Some of the ordered pairs, comparing depth and psi, would be (0, 14.7), (33, 29.4), (66, 44.1), and so on. After selecting two ordered pairs, the slope could be found: $\frac{29.4-14.7}{33-0} = \frac{14.7}{33} \approx 0.45$. Since the y-intercept is at 14.7, an equation relating depth and pressure is P = 0.45d + 14.7.

Using the previous example as a guide, students could work out a similar problem whose data are in different units. One atmosphere is defined as the pressure at sea level, which is 14.7 psi. Also, 33 feet is approximately 10.06 meters. Thus, every descent of 10.06 meters is 1 additional atmosphere of pressure. Thus another set of ordered pairs, this time relating meters and atmospheres, is (0,1), (10.06,2), (20.12,3), etc.

Boiling Point vs. Elevation

At sea level, the boiling point of water is 212 degrees Fahrenheit. However, at higher elevations, water boils at a lower temperature. Changes in the elevation, and thus the density of the air will cause that boiling point to decrease as the altitude increases. The following table relates altitude and the temperature at which water boils.[27]

Altitude (feet)	Temperature (°F)
0	212
500	211.1
1000	210.2
2000	208.3
5000	202.8

Students could find various slopes using these data points to verify that the relationship seems to be linear. Then using two points, students could find an equation relating the two variables.

This boiling point issue can be examined in a different way. Although related to the previous example, in this case, the information might be presented in the form of a graph. The graph seems to be linear. The information presented in the graph can again be written as a linear equation comparing elevation to the boiling point of water.

This could be done a couple of ways. One would be by using the point-slope formula. Two ordered pairs could be selected from the graph. Using those points, the slope could be found. Then one of the two points and that slope would be substituted into the point-slope formula.

Another way could be to use the slope-intercept form. Since (0,212) is a point on the graph, 212 is the value of the y-intercept. The slope could be estimated by counting spaces from one point to another and finding the rise over the run. A student might plausibly state that the graph seems to drop 2 for every run of 6. However, students would have to examine the scales of the axes. What looks to be a drop of two spaces might actually be a drop of 10 and a run of 6 spaces is actually a run of 6000. So the slope in this case would be –10 over 6000. We could then use the slope-intercept form with $m = -\frac{1}{600}$ and b = 212. So we have the equation $y = -\frac{1}{600}x + 212$.

On a graph it is sometimes difficult to tell exact values for points, so student answers may vary somewhat. For the sake of comparison, it might be helpful to know that the actual formula that relates elevation to boiling point is y = –0.00184x + 212.[28]

Depreciation

Depreciation is the decrease in value of an asset. There are different ways a company can figure depreciation. One of these is called straight-line depreciation. As can be guessed from its name, this is a linear relationship.

Examples:
1. Suppose a company's assets of $40,000 depreciate at $3,000 per year.

Thus, after year one, those assets are worth $37,000. After year two, they are worth $34,000, and so on. This straight line relationship would have its beginning (the y-intercept) at $40,000 and decrease $3,000 every year thereafter (a slope of –3,000). An equation representing its depreciation over time would be y = –3,000x + 40,000.

2. A $50,000 company car is worth $36,000 after four years. What is an equation that expresses this depreciation relationship?

The slope is $\frac{36,000-50,000}{4} = -3,500$. Thus, the equation is y = –3,500x + 50,000.

3. A car is worth $20,000 after three years and worth $10,000 after five years. What is the rate of depreciation, the original price of the car, and an equation showing the depreciation over time?

The slope is $-\frac{20,000-10,000}{3-5} = -5,000$, which is the rate of depreciation. So, y = –5,000x + b. Substituting the point (3, 20,000), gives the value of b (the original cost), $35,000. The depreciation equation is y = –5,000x + 35,000.

Though not linear, there are other types of depreciation. A double declining balance method is an exponential relationship. With this method, the reduction will be more in the early years of the asset's life and less in the later years.

Above, there was a $50,000 car that was declining at $4,000, or 8 percent, every year. With a double declining balance method, that percentage is doubled to be 16 percent, and is multiplied each year by the remaining value of the car to find the amount of depreciation. Thus, after one year, our $50,000 car is figured to be worth $42,000. The next year it is worth $35,280. An equation expressing this relationship is y = 50,000(0.84^x).

Line of Best Fit

A line of best fit, also known as a regression line, is a line that best represents a set of ordered pairs. There are a number of ways students can find these lines.

One way is to have students plot those ordered pairs, and, with a straightedge, draw in what looks to be the best fitting line. From that line, students could select two representative points and use those to find the slope and then write an equation for the line. That equation could then be used to make predictions regarding future values. Going through this process may not give the absolute *best* fitting line, but it does allow students to have a good understanding of the line of best fit concept.

Alternately, students could use the least squares method to find the equation by hand (shown below). This will produce a correct answer, but is a lot of work. The easiest way is to input the ordered pair data into a calculator or computer program made for such jobs.

There are a number of sets of data that can be used that have a nearly linear relationship. The decline of cigarette consumption over the years, the increase of life expectancy, and the relationship between football players' height and weight are just a few of many examples.

There are also cases that show the danger of using these lines without putting in some common sense thought. For example, a line of best fit can be found that compares children ages 5 to 16 and their heights. Using that line of best fit, which might work well for children, would give a strange result when using its equation to find the projected height of a 40-year-old.

Data could be plotted that shows the winning time in the Olympic Games in the 1500-meter run. The winning times have gotten faster through the years and, when plotted on a graph, might look to be a linear relationship. However, making this assumption leads to the conclusion that, at some point, someone will win the Olympic 1500-meter run in zero seconds.

Evidence has been collected that shows that the Earth is heating up. A linear equation could be found representing this data. But is this line a reliable one for predicting the future? Some would say that unless something is done, the Earth will continue to heat up as predicted by this linear model. Others might say the Earth's temperature is cyclical and we are only seeing a small part of its history. They would say that looking at the big picture, our line of best fit should really be a sine wave. It could make for an interesting class discussion.

As promised, this is the method of least squares. WARNING—This is not for the faint of heart. However, doing things the longhand way can give a student an awareness of how things were done previously and a sense of thankfulness for current technology. The following expressions find the slope and y-intercept in the regression line, y = mx + b, and the correlation coefficient, r. The x's and y's stand for the (x,y) ordered pairs, and n is the total number of ordered pairs.

$$m = \frac{n\sum xy - \sum x \sum y}{n\sum x^2 - (\sum x)^2} \quad b = \frac{n\sum x^2 \sum y - \sum x \sum xy}{\sum n \sum x^2 - (\sum x)^2} \quad r = \frac{\sum xy - \sum x \sum y}{\sqrt{n(\sum x^2 - (\sum x)^2}\sqrt{n(\sum y^2 - (\sum y)^2}}$$

Example:

While this would typically be used for more than a set of three points, let's do so in this case to simplify calculations. Using the points (1,1), (4,1), and (5,0) we can find m and b. If seen through to completion, we find m = –0.192 and b = 1.308. Thus, the line of best fit is y = –0.192x + 1.308. Plotting those points show that the line does come reasonably close to the points, yet isn't a perfect fit. Values of r range from –1 to 1, with –1 and 1 showing a perfect correlation and 0 showing that the points have no correlation. For this regression line, the correlation turns out to be –0.6934. This would imply that as the x-values increase, the y-values decrease, though not in a perfectly linear pattern.

Solving First Degree Equations

Solving Percent Problems

What percent is 17 of 32? Problems like this often perplex students. "Do I multiply 17 and 32? Do I divide them? If I do divide, what goes into what?" While there are different ways to approach these problems, a basic knowledge of how to solve first degree equations can make those decisions easy for students. A student can set up an equation by learning some simple translations ("is" corresponds to an equal sign, "of" corresponds to multiplication) and then solving the equation.

Examples:
1. 17 is what percent of 32? → $17 = x\% \cdot 32$
2. What is 14 percent of 85 → $x = 0.14 \cdot 85$
3. 108 percent of what is 54 → $1.08 \cdot x = 54$

Distance = Rate x Time

Some students seem to have a good feel for distance/rate/time problems. However, it can be a guessing game for many students until learning basic principles of algebra. Similar to the above percent problems, algebra and the formula $d = r \cdot t$ take away the guesswork of what operations to use.

Examples:
1. You are going on a 312-mile trip. Averaging 57 miles per hour, how long would the trip take?

$$312 = 57 \cdot x \quad \text{[Answer: 5.474 hours, or 5 hours, 28 minutes]}$$

2. The Sun is approximately 93 million miles from the Earth. The speed of light is 186,000 miles per second. How long would it take light from the Sun to reach us?

$$93{,}000{,}000 = x \cdot 186{,}000$$
[Answer: 500 seconds (8 minutes, 20 seconds)]

3. If a laser from Earth strikes the Moon in 1.274 seconds, how far apart are they?

$$x = 186{,}000 \cdot 1.274 \quad \text{[Answer: 236,964 miles]}$$

4. It is often stated that a person is a mile away from a storm for every five seconds that pass between lightning and thunder. Is this true? The $d = r \cdot t$ formula can be used to obtain the time it would take for the sound to

travel. Sound travels at roughly 761.2 miles per hour. Solving 1 = 761.2t shows that it takes 0.001314 hours for the sound to travel a mile. Multiplying 0.001314 hours by the number of seconds in an hour (3,600) gives 4.73 seconds. So 5 seconds is a pretty good estimate.

Ohm's Law

In 1827, Georg Ohm, for whom the unit of resistance is named, found the relationship between these three basic units of electricity.

$V = I \cdot R$,   where I = Current (measured in amperes or amps)
            V = Voltage (in volts)
            R = Resistance (in ohms)

The voltage is the difference in charge between two points on a circuit. The current is how fast the charge is flowing. The resistance is the tendency to resist that flow.

The flow of electrons through a wire is something like the flow of water in a river. Continuing with the water analogy, the voltage might be the difference in elevation between the river's source and where it ends. The current is how fast the river is flowing. The resistance might be rocks and bends in the river that could slow things down.

Examples:
1. A circuit contains a voltage of 3.2 volts and resistance of 2 ohms. How much current flows in the circuit?

$$V = I \cdot R \rightarrow 3.2 = I \cdot 2 \rightarrow I = 1.6 \text{ amps}$$

2. A circuit has a current of 3.6 amps and a resistance of 4.1 ohms. What is the voltage? [Answer: 14.76 volts]

3. A 12-volt source draws 4.8 amps of current. How much resistance is contained in this circuit? [Answer: 2.5 ohms]

Boiling Point

Water boils at 100 degrees Celsius. At least it does at sea level. A general rule of thumb is that the boiling point drops five degrees for every mile gained in elevation. The equation P = 100–5m expresses this relationship, where P is the boiling point in degrees Celsius and m is the number of miles above sea level. (We previously examined the effect elevation has on water's boiling point, but that was using the Fahrenheit scale and the elevation in feet. Many of these applications can be reused by making changes in the measurement units.)

Examples:

1. Mount Everest is about 5.5 miles high. What would the boiling point of water be on top of Everest? [Answer: 72.5 degrees]

2. The Dead Sea is about a quarter of a mile below sea level. What is the boiling point there? [Answer: 101.25 degrees]

3. How many miles are you above sea level if water boils at 87 degrees Celsius? [Answer: 2.6 miles]

Simple Interest

Simple interest does not involve compounding. Pawn shops or others dealing with short term loans would be apt to use the simple interest formula. Banks and credit card companies use compound interest. The formula for simple interest is I = prt, or interest equals principal times rate times time. A related formula is A = p(1 + rt). The first formula would give only the interest you earn (or, in the case of a loan, how much you would have to pay in interest). The second formula computes the interest earned added back onto the original money. For the following examples, we'll assume simple interest is being used.

Examples:

1. Retirement planners often use what is called the four percent rule. The concept is that, upon retirement, an individual should draw out no more than 4 percent each year of what is currently saved. Your money should last as 4 percent is the commonly assumed annual growth rate for investments. What is an estimate of how much should be saved for retirement if you want to be able to take out $4,500 per month? This could be found by solving $4{,}500 = x \cdot 0.04 \cdot \frac{1}{12}$. [Answer: $1,350,000]

2. You notice on your credit card bill that last month's charges of $723 caused you to have $11.15 in interest charges. What percent interest are you being charged?

$$11.15 = 723 \cdot x \cdot \tfrac{1}{12} \quad \text{[Answer: 18.5 percent]}$$

3. A friend asks to borrow $150 and says he will pay you back $200 in 6 months. If you agree to this, what amount of interest would you be earning on this investment?

$$200 = 180(1 + 0.5r) \quad \text{[Answer: 22.2 percent]}$$

Hooke's Law

In the 1600s, English scientist Robert Hooke discovered a relationship between the force used in compressing (or stretching) a spring and the distance it is compressed. That formula is F = kx, with F being measured in newtons and x in meters. The constant k will vary for different springs as some are "springier" than others. (Hooke's Law is sometimes written as F = – kx since the spring is exerting a force in the opposite direction to get back to its equilibrium position.)

Example:
A force of 7.4 newtons compresses a spring 2 meters. How far would a force of 8.3 newtons compress the spring?

This is a two-part problem. First k would need to be found to create the equation that will be used to solve the problem.

$$F = kx \rightarrow 7.4 = k \cdot 2 \rightarrow k = 3.7$$

The equation F = 3.7x now can be used for any problem involving this particular spring. In this case, to find the distance a force of 8.3 newtons would compress this spring:

$$8.3 = 3.7x \rightarrow x = 2.24 \text{ meters}$$

Center of Mass on a Number Line

Suppose there are two objects located on a number line. The first object is located at x_1 and has a mass of m_1. A second object located at x_2 has mass m_2. The center of mass is located at $C = \frac{m_1 x_1 + m_2 x_2}{m_1 + m_2}$.

Examples:
1. Object A, weighing 20 pounds, is located at 5 on a number line. Object B weighs 100 pounds and is located at 30. What is the location of the center of mass?

$$C = \frac{20 \cdot 5 + 100 \cdot 30}{20 + 100} = 25\frac{5}{6}$$

This answer would place the center of mass much closer to object B then object A. This makes sense since object B is much heavier than object A.

2. From a given point, a 23 pound object is located at –7. A second object weighs 8 pounds. The center of mass is at 3. What must be the location of that second object?

$$3 = \frac{23 \cdot (-7) + 8 \cdot x}{20 + x} \rightarrow 60 + 3x = -161 + 8x \rightarrow 221 = 5x \rightarrow x = 44.2$$

Similarly, the center of mass in two dimensional cases can be examined. Consider three points (x_1, y_1), (x_2, y_2), and (x_3, y_3), having respective masses of m_1, m_2, and m_3. The x-value and y-value of the center of mass are found individually.

$$x = \frac{m_1 x_1 + m_2 x_2 + m_3 x_3}{m_1 + m_2 + m_3} \quad \text{and} \quad y = \frac{m_1 y_1 + m_2 y_2 + m_3 y_3}{m_1 + m_2 + m_3}$$

Weighted Averages

Algebra could be used to find the missing value needed to yield a specific mean average.

Example:
The final grade for a course is based on the scores of five tests given during the grading period. A student has taken four of the tests so far, with scores of 92, 83, 76, and 77. What score is needed on the final test to get a 70 percent? An 80 percent?

$$\frac{92 + 83 + 76 + 77 + x}{5} = 70 \qquad \frac{92 + 83 + 76 + 77 + x}{5} = 80$$

Our hypothetical student would find out that getting a C in the course means scoring at least a 22 percent in the final test, which should be no problem. However, our student would also find out that getting a B means needing to get at least a 72 percent on the final test.

Weighted averages are similar to the mean average. While weighted averages are seen in a number of different areas, students are likely most familiar with this concept as used by teachers to determine grades. Teachers could choose to attach different "weights" to various parts of their grading scale.

Examples:
1. A teacher has a grading system in which 20 percent of a student's grade will be based on assignments, 30 percent on tests, and 50 percent on a final project.

What is the grade for a student that has scored 79 percent on assignments, 53 percent on tests, and 82 percent on the final project? To find that student's grade means computing the weighted average:

$$x = 0.20(0.79) + 0.30(0.53) + 0.50(0.82) = 0.727$$

With a score of 72.7 percent, the student will receive a grade of C.

2. Another student has a goal of getting a B, which means obtaining at least an 80 percent. She knows already that she has total assignment grades of 78 percent and tests of 73 percent. All she has left is to complete the final project. She wants to know what grade she needs on the project to get her B. The answer lies in solving the equation:

$$0.20(0.78) + 0.30(0.73) + 0.50(x) = 0.80 \rightarrow x = 0.8482 \approx 85\%$$

Grade Point Average

Speaking of grades, an important number for many students is their grade point average or GPA. Most students have a general feel for what a

GPA signifies. They might know a 3.43 means you get mostly A's and B's. However, many don't seem to know how to calculate their GPA. A grade point average is really a weighted average in which A's are worth 4 points, B's worth 3, C's worth 2, and D's worth 1. F's, as should be, aren't worth anything. The grade point average can be found by substituting the numbers of A's, B's, C's, D's, and F's into the following formula:

$$GPA = \frac{4A + 3B + 2C + 1D + 0F}{7}$$

Examples:

1. A student received grades that consist of 2 A's, 3 C's, a D, and an F. Those 7 grades would yield a GPA of:

$$GPA = \frac{4(2) + 3(0) + 2(3) + 1(1) + 0(1)}{7} = 2.14$$

2. Suppose a student would like to have a GPA of 3.0. He knows most of his grades already. He has an A, 3 B's, a C, and a D so far. He's not sure about his remaining class. What would he need?

$$\frac{4(1) + 3(3) + 2(1) + 1(1) + x}{7} = 3.0$$

It turns out that x would equal 5. Sadly, that isn't going to be possible in this grading system.

Momentum

To find an object's momentum, mass is multiplied by velocity. Any football player could tell you that if two players traveling at the same speed hit you, the 300-pound player has a much greater impact than the 160 pounder. Emergency escape ramps are built adjacent to mountainous roads because large trucks' brakes might not be able to slow them down. The conservation of momentum principal states that although masses and velocities might change, the total momentum in a system stays the same. This would be true for collisions involving football players, automobiles, or atoms. If two objects, with masses m_1 and m_2, lie along a straight line and are travelling with velocities v_1 and v_2, their new velocities, v_3 and v_4, will be such that momentum will be conserved. Thus:

$$m_1v_1 + m_2v_2 = m_1v_3 + m_2v_4$$

Examples:

1. Object A, with a mass of 8 kilograms and moving at 5 meters per second, hits object B which has a mass of 6 kilograms and is moving in the same direction at 1 meter per second. After the collision, object B is now

moving at 10 meters per second. How fast is object A going after this collision takes place?

$$8 \cdot 5 + 6 \cdot 1 = 8x + 6 \cdot 10 \to 46 = 8x + 60 \to x = -1.75 \text{ m/sec}$$

The fact that the solution is negative means object A bounced off object B and is now moving in the opposite direction.

2. A 5 kilogram gun is going to fire a 0.006 kilogram bullet. The bullet leaves the gun at 800 meters per second. If not braced against anything, how much would the gun recoil?

Keeping in mind that the gun and the bullet are both motionless before the firing, the equation is $5 \cdot 0 + 0.006 \cdot 0 = 5v + 0.006 \cdot 800$. Again, the solution will be a negative number as the gun will go in the opposite direction from the bullet. [Answer: –0.96 m/sec]

3. A 250-pound lineman moving to the right at 11 miles per hour hits a 210 pound running back coming toward him at 9 miles per hour from the opposite direction. After the collision, the lineman continues to move to the right, but now at two miles per hour. How fast was the running back moving after the hit?

Taking movement to the right to be in the positive direction, there is the following formula:

$$250(11) + 210(-9) = 250(2) + 210x \to x = 1.714$$

This second person is moving at roughly 1.7 miles per hour. Because it is a positive number, he must be moving to the right, the opposite way he was running.

Mixture Problems

Mixtures are related to the concept of weighted averages. Two or more substances can be combined to form a mixture. Here are three examples.

Examples:
1. There are two bottles that contain 4 percent and 16 percent acid solutions. Two liters of a 7 percent acid solution is needed. How much of each bottle should be mixed together to obtain this solution?

Two liters are needed, so if x liters of the first bottle are used, there are 2 – x liters needed from the second bottle. The equation can be written:

$$0.04x + 0.16(2 - x) = 0.07(2)$$

Solving this equation gives us an answer of x = 1.5. So, one and a half liters of the 4 percent solution and a half liter of the 16 percent solution are needed.

2. A radiator currently contains 4 gallons of a solution that is 35 percent antifreeze. However, you are advised that it would be better if it were at least

a 40 percent antifreeze solution. How much more pure antifreeze would have to be added to bring it to the 40 percent amount?

The situation is that 4 gallons of 35 percent antifreeze plus an unknown amount of pure (100 percent) antifreeze equals the final mixture. This results in the following equation:

$$(0.35)4 + (1.00)x = (0.4)(4 + x)$$
[Answer: x = ⅓ of a gallon of pure antifreeze]

3. Your hot tub currently has 400 gallons of water at 95 degrees. The water level is a bit low, so you must add some water from your garden hose. You estimate you are putting in about 20 gallons of 40-degree water. What is the new temperature of the hot tub going to be? And is this going to be warm enough for you to get in right away?

This time there is a mixture of temperatures, but the set up to the equation will be similar to the other mixture problems.

$$95(400) + 40(20) = x(420)$$ [Answers: x = 92.4° and probably]

Expected Value

Expected value is the average value of what one would expect to receive in repeated trials. It is computed by finding the sum all possible occurrences each multiplied by the probability they occur. Suppose a Mr. Shyster invites you to play a game of chance with him. He rolls a die. If it comes up a one, you win $7. If it comes up any other number, you have to pay him $2. Should you play? In the long run, the average roll would have a value of $\frac{1}{6} \cdot 7 + \frac{5}{6} \cdot (-2) = -\frac{1}{2}$. Although no single roll would cost this amount; on the average, you would lose fifty cents a turn.

Examples:

1. If the goal is to have a balanced game, how much should you have to pay Mr. Shyster each time you lose? The answer would be found with the equation $\frac{1}{6} \cdot 7 + \frac{5}{6} \cdot x = 0$. The answer turns out to be $-\frac{7}{5}$, or, in other words, you should only pay $1.40 for each loss to make the game fair.

2. You buy tickets to a concert for $60. The ticketing agency has insurance in the event the performance is cancelled. For the insurance of $14, the buyer is refunded if there is a cancellation. Is this a good deal?

(A.) If you don't buy the insurance: Your cost is $60, whether there is a show or not.

(B.) If you get the insurance: If there is a show, you are out the $60 in tickets and $14 in insurance. If there is a cancellation, you are only out $14.

Concerts usually don't get cancelled. Let's assumed there is a 10 percent chance it will be cancelled. With option A, your cost is $60 regardless what happens. With option B, your expected cost is $74 \cdot 0.9 + 14 \cdot 0.1 = \$68$. That doesn't beat the $60 cost without insurance, which is $60. So, assuming a 10 percent probability of cancellation, you would be better off without buying the insurance.

At what percent chance of cancellation would insurance make sense? Let x be the probability of cancellation. Then $1 - x$ would be the chance the show goes on.

$$60 = 74(1 - x) \to 60 = 74-60x \to x = .2\overline{3}$$

If your performer is on the sickly or flaky side and cancels more than 23.3 percent of the time, you would want to get the insurance. Otherwise, no. (These were actual numbers from a recent purchase. The experts say that product insurance is generally not a good idea for the consumer. This example shows that to be the case.)

Average Atomic Mass

A particular element always has the same number of protons. However, there can be various isotopes of that element depending on the number of neutrons it possesses. The masses of the isotopes, along with how often they occur, determine the element's average atomic mass.

Examples:
1. Silver has two isotopes; one with an atomic mass of 106.90509 and another with an atomic mass of 108.90470. If the atomic mass were based on a simple mean average, the numbers would be added and divided by two. However, these two isotopes occur in nature 51.86 percent and 48.14 percent of the time, respectively. Thus, to find the average atomic mass:

$$106.90509(0.5186) + 108.90470(0.4814) = 107.868$$

That value, 107.868, can be found on the periodic table under Ag, the symbol for silver.

2. The periodic table states that carbon has an average atomic mass of 12.01115. Carbon has two naturally occurring isotopes. The most common is carbon-12 with an atomic mass of exactly 12. Its abundance is 98.9 percent. The other is carbon-13. What is its carbon-13's atomic mass?

Simple arithmetic tells us that the occurrence of carbon-13 must be $(100\% - 98.9\% =)$ 1.1%. Now, algebra is used to solve:

$$12(0.989) + x(0.011) = 12.01115 \text{ [Answer: } x = 13.014]$$

Speed of Sound vs. Temperature

A major factor in how fast sound travels is the temperature of the medium the sound is traveling through. A good approximation for the speed of sound as it travels through air is v = 331.4 + 0.6t, where v is the speed of sound in meters per second and t is the temperature in degrees Celsius.

Example:
At what temperature would sound travel at 350 meters per second?
The equation 350 = 331.4 + 0.6t gives a temperature of 31 degrees Celsius.

Octane Numbers

The expression (R + M)/2 is often seen on gas pumps, and is used to obtain the octane rating for gasoline. Gas pumps also will have printed various ratings of gasoline such as 87, 89, etc. These are obtained by using this average. Without getting into too much detail, R is the Research Octane Number and M is the Motor Octane Number. These values are determined by running the fuel in a test engine. R is a value obtained while the engine is running at 600 rpm and M is a value when running at 900 rpm and using preheated fuel.[29]

There are many different types of fuel. For example, m-butanol has an R value of 92 and an M value of 71. The (R + M)/2 formula could be used to find the octane rating. Also, when given the overall octane rating and the R or M value, the other value could be found using algebra.

Fahrenheit/Celsius Conversions

The formula that converts a temperature in Celsius to Fahrenheit is $F = \frac{9}{5}C + 32$.

Example:
The normal body temperature is 98.6 degrees Fahrenheit. What is the normal body temperature in degrees Celsius?

$$98.6 = \frac{9}{5}x + 32 \quad \rightarrow \quad 66.6 = \frac{9}{5}x \quad \rightarrow \quad x = \frac{5}{9}(66.6) = 37.0 \text{ degrees Celsius}$$

Baseball Statistics

There are many formulas used in computing baseball statistics. A good application is finding a baseball player's batting average. The batting average is the ratio of hits to official at-bats. An at-bat that is not "official" would refer to getting on base by means other than batting ability. For example, being walked or being hit by a pitch is not counted. Batting average is expressed as a decimal rounded to thousandths place. A person getting one hit in four at-bats is hitting 0.250, pronounced "two fifty." A person going two for three is batting 0.667, pronounced "six sixty-seven." Do not get mathematically correct and pronounce this "six hundred sixty-seven thousandths." Expect blank stares or ridicule if you do. Incidentally, when baseball people are talking about Ted Williams being the last person to "hit four hundred," they mean the last person to have a hit to at-bat ratio being greater than or equal to 0.400.

Here are a few of the formulas used to judge a batter's ability:

$$\text{Batting Average (BA)} = \frac{H}{AB}$$

$$\text{Total Bases (TB)} = S + 2D + 3T + 4HR$$

$$\text{Slugging Average (SA)} = \frac{TB}{AB}$$

For these formulas, H = Hits, AB = Times at bat, S = Singles, D = Doubles, T = Triples, and HR = Home runs.

Example:
The following table makes for an interesting exercise. Below are a season's statistics for four major league players.

Player	AB	H	S	D	T	HR	TB	BA	SA
Mike Trout	602	173	86	39	9	36	338	.287	.549
Pete Rose	662	210	152	47	4	7	286	.317	.432
Willie Mays	600	208	135	33	11	29	350	.347	.583
Babe Ruth	529	200	108	39	7	46	391	.378	.739

By taking actual player's statistics while leaving certain categories blank, students would face a mixture of problems to be solved with logic and the above formulas. It might look like this:

Player	AB	H	S	D	T	HR	TB	BA	SA
Mike Trout	602		86	39	9	36		.287	
Pete Rose	662	210	152	47		7	286		.432
Willie Mays	600	208	135	33	11	29	350	.347	.583
Babe Ruth		200	108	39	7		391	.378	

More or fewer blanks could be inserted to make the exercise more or less challenging.

Pitchers also have statistics associated with what they do. The earned run average (ERA) is:

$$\text{ERA} = \frac{9E}{I}, \text{ where E = Earned runs allowed and I = Innings pitched.}$$

This formula gives the average number of earned runs that would score against the pitcher in a complete nine inning game. Earned runs are those that are the fault of the pitcher. For example, they do not include runs that are scored due to fielding errors. This is an interesting formula as it often contains fractions. A pitcher that has gone 4 innings and has gotten one out in the next inning is considered to have pitched 4⅓ innings. Getting two outs is counted as 4⅔ innings pitched.

Example:

A player's ERA is 3.24 after giving up 3 earned runs in his last game. How many innings did he pitch?

$$3.24 = \frac{9(3)}{x} \rightarrow 3.24x = 27 \rightarrow x = 8\tfrac{1}{3} \text{ innings pitched.}$$

Chirping Crickets

Crickets communicate by chirping. Students may know, either from their own experience, or from being featured on an episode of *The Big Bang Theory,* that crickets chirp at a faster rate as the temperature warms. This fact was discovered by Professor Amos Dolbear at Tufts College in 1897. He not only discovered that different types of crickets will chirp at different rates, but developed formulas for various types of crickets.

In the following formulas, N = number of chirps per minute, and T is the temperature in degrees Fahrenheit.[30]

Field Cricket: $\qquad T = 50 + \frac{N-40}{4}$

Snowy Tree Cricket: $\qquad T = 50 + \frac{N-42}{4.7}$

Katydid: $\qquad T = 60 + \frac{N-19}{3}$

Examples:

These formulas could lead to a number of questions. Some possibilities:

1. If a katydid chirps 13 times in a minute, what is the temperature?

2. If the temperature is 72 degrees, how many chirps would you expect a snowy field cricket to make in a minute?

3. It is known that crickets rarely make noise below 55 or above 100 degrees Fahrenheit. If so, what is the domain and range of the snowy tree cricket function?

4. At what temperature would the field cricket and the katydid have the same number of chirps per minute?

5. By comparing their slopes, which cricket's chirp rate increases the fastest?

6. Since the relationship between the Fahrenheit and Celsius scales is $F = \frac{9}{5}C + 32$, compose this function with the katydid formula to find the relationship between its chirps per minute and the temperature in degrees Celsius.

Tax Rates

For the 2014 tax year, one of the federal tax brackets has this to say for those that are married and filing jointly: "If your taxable income is over

$73,800, but not over $148,850, your tax is $10,162.50, plus 25 percent of the excess over $73,800."

Examples:
1. Write this tax instruction as a function.

[Answer: f(x) = 10,162.50 + 0.25(x − 73,800), for 73,800 < x ≤ 148,850]

2. If your federal taxes came to $11,712, what was your taxable income?

11,712 = 10,162.50 + 0.25(x − 73,800) → x = $79,998

3. What are the amounts of tax at the lower and upper ends of this tax bracket?

[Answer: $10,162.50 and $28,925]

Inequalities

There are inequality relationships to be found in many areas. If your systolic pressure is less than 90 (x < 90), you may have blood pressure that is a bit too low. Well trained athletes might have resting heart rates in the 40 to 60 beats per minute range (40 ≤ x ≤ 60). The Pew Research Center defines middle income as being between two-thirds and double the median income. For a three-person household, that amount is considered to be between $42,000 and $126,000 (42,000 ≤ x ≤ 126,000).[31]

Many of the examples of equations in this book can easily be converted to inequalities. In fact, many real world situations make more sense as inequalities. A person may want to know how many hours she has to work to make $5,000 monthly. More likely, what she really wants to know is how many hours she has to work to make *at least* $5,000. Following are a few examples of applications that contain inequalities.

1. A softball player wants to end the season with a batting average (hits divided by at-bats) of at least .300. With one day left in the season, she just barely makes it with 145 hits in 483 at-bats. She figures she'll get up to bat four times in her final game. How many hits does she need to hit .300?

$\frac{145 + x}{487} \geq 0.300$ [Answer: x ≥ 1.1 (so she better plan to get at least 2 hits)]

2. With a long trip ahead, you plan to drive 55 miles per hour and want to make between 250 to 300 miles per day. How many hours do you need to drive each day to achieve this?

250 ≤ 55x ≤ 300 [Answer: Between 4.55 and 5.45 hours each day]

3. In a college class, the results from three tests will determine a person's

grade. You have taken two tests and scored 71 percent and 83 percent. What is needed on the third test to end up with a B?

$$\frac{71 + 83 + x}{3} \geq 80 \quad \text{[Answer: At least an 86 percent]}$$

4. A used car salesman is guaranteed $500 per month along with an 8 percent commission on all of his sales. If a particular car salesman would like to have a monthly salary of at least $6,000, how much in car sales does he need to achieve this?

$$500 + 0.08x \geq 6,000 \text{ [Answer: At least } \$68,750]$$

5. We saw in the section on absolute value that tolerances could be written as inequalities containing absolute value. For example, an object is to be manufactured with a length of 7.3 meters and a tolerance of ±0.02. This can be translated to $|7.3 - x| \leq 0.02$. This inequality can be solved by breaking the inequality into:

$$7.3 - x \leq 0.02 \text{ and } 7.3 - x \geq -0.02$$
[Answer: Between 7.298 and 7.302]

Mean Arterial Pressure

A blood pressure reading of 120/80, or "120 over 80" is the ratio of the systolic pressure (S), (the arterial pressure when the heart beats) to the diastolic pressure (D), (the arterial pressure between beats). The mean arterial pressure (MAP) gives a single reading. Since the diastolic phase lasts twice as long, the formula looks like this[32]:

$$\text{MAP} = \frac{2D + S}{3} = \frac{2 \cdot 80 + 120}{3}$$

A reading of 120/80 would be a MAP of 93⅓.

It is believed that a MAP of at least 60 is necessary to sustain the health of organs of the body. This being the case, what would the relationship be between the systolic and diastolic pressures? An inequality to express this relationship is:

$$60 \leq \frac{2D + S}{3}$$

$$180 \leq 2D + S$$

$$S \geq -2D + 180$$

This inequality could be written as $y \geq -2x + 180$ and graphed fairly easily on a coordinate plane. This graph would show which of the various pressures are life-supporting and which are not.

Television Viewing Distance

The website of a television manufacturer has the following statement: "To find the best viewing distance we recommend starting with industry standards, which call for multiplying the TV screen size by 1.2 (minimum distance) and 1.9 (maximum distance). Then divide the result by 12 to get the right number of feet."[33]

Those English sentences could be translated into mathematics with the compound inequality $\frac{1.2x}{12} \le y \le \frac{1.9x}{12}$, where x is the television screen size in inches and y is the viewing distance in feet. Given a value, x, for the size of a television screen, the range in feet one could sit from the screen to get optimal viewing pleasure could be found. On the other hand, if the distance from screen to viewing location is already known, the range of optimal television sizes could be found.

Examples:
1. What are good distances to sit from a 54-inch television?

$$\frac{1.2(54)}{12} \le y \le \frac{1.9(54)}{12} \quad \rightarrow \quad 5.4 \le y \le 8.55$$

A good viewing distance from this television screen would be sitting roughly between 5½ and 8½ feet from the screen.

2. You plan to purchase a television that will be positioned 9 feet from your living room couch. What size television should you buy?

$$\frac{1.2x}{12} \le 9 \le \frac{1.9x}{12} \quad \rightarrow \quad \frac{1.2x}{12} \le 9 \text{ and } 9 \le \frac{1.9x}{12} \quad \rightarrow \quad x \le 90 \text{ and } x \ge 56.8$$

So, there are a range of possible television screens from 56.8 to 90 inches.

Linear Programming

A linear programming problem is a good application of graphing linear inequalities. Linear programming would be used in a situation in which there is a need to find an optimal solution. However, as is true in much of life, there may be constraints which limit the possibilities. Those constraints are written as linear inequalities. When graphed, those inequalities often form a polygon. The vertices of the polygon become the candidates for the optimal solution. Those points are then substituted into the original function to determine which of those possibilities is indeed best.

Examples:
1. A supermarket might think there should be as many checkout locations as possible. If there are long lines, customers might decide it is better to shop at another store. However, if there are too many, the store might end

up paying checkers that spend much of the time just standing at their checkout stations waiting for customers. What is the optimal number of checkout locations to maximize profit?

    2. A factory makes money by manufacturing stereos. How many should it produce? While one might think, "As many as possible," there may be limits to what is possible. What the factory machinery can produce, the number of workers you have, the storage space available, and the ability to ship the items are all possible constraints on production.

Systems of Equations

Equal Temperatures

    A temperature of 100 degrees Celsius corresponds to 212 degrees Fahrenheit. To subtract the two numbers gives a difference of 112. We also know that 0 degrees Celsius corresponds to 32 degrees Fahrenheit. Now there is a difference of only 32. It would seem that the colder it gets, the less difference there is between the two scales. Is there perhaps a point at which the Fahrenheit and Celsius scales read the same number? It turns out there is. It can be found by solving the system of equations:

$$y = \tfrac{5}{9}x + 32 \quad \text{and} \quad y = x$$

Students will probably see that the easiest way to approach this problem would be with the substitution method, although it would be an interesting exercise to graph the two as well. Either method gives a temperature of 40 degrees below zero.

Shooting Percentages

    A basketball player's free throw percentage is found by the number of free throws made, divided by the number of attempts. For a given free throw percentage, there is not enough information to know how many actual free throws were made or attempted. A free throw percentage of 50 percent could be the result of a player going 3 for 6 or 50 for 100. With a little more information, though, two equations could be set up and solved.

    Example:

    1. Leina keeps track of her shooting percentages. She knew her free throw percentage was 60.9 percent. During her next game she made 6 out of 7 free throws and was told her free throw percentage had risen to 66.7 percent. How many free throws had Leina attempted and made?

$$\frac{x}{y} = 0.609 \quad \text{and} \quad \frac{x+6}{y+7} = 0.667$$

The first equation can be rewritten as x = 0.609y. Making a substitution into the second equation gives:

$$\frac{0.609y + 6}{y + 7} = 0.667 \quad \rightarrow \quad x = 13.98 \text{ and } y = 22.95$$

One would expect whole number answers. The shooting percentages apparently were rounded values. So the answer to the problem must be that Leina made 14 out of 23 free throws, and after her 6 for 7 game, she had made 20 out of 30.

While on the subject of basketball, the number of points scored in a basketball game can be found with the equation P = x + 2y + 3z, where x is the number of free throws made, y is the number of two-point field goals, and z is the number of three-point field goals. Suppose a player reads his stat line and knows he scored 27 points by making 6 free throws and 10 field goals. How could you determine the number of two-point and three-point field goals?

Two equations could be set up: 27 = 6 + 2y + 3z and y + z = 10
[Answer: 9 two-point field goals and 1 three-point field goal]

Students might have been able to find the previous answers by a guess and check method. While that is commendable, a teacher might want to force the issue of solving with other methods by making the numbers much larger. The guess and check strategy then proves to be much tougher. For example, in his career, Michael Jordan made 7,327 free throws and 12,192 field goals for a total of 32,292 points. The field goals were a mixture of two-pointers and three-pointers. How many were there of each?

This calls for solving the system of equations:

32,292 = 7,327 + 2y + 3z and 12,192 = y + z.
[Answer: 11,611 two-point field goals and 581 three-point field goals]

Equations of Conics

The path taken by a projectile is a parabola. The equation of a parabola can be written $y = ax^2 + bx + c$. In order to write an equation for a specific parabola, the values for a, b, and c would need to be found. Just as it takes two points to determine a line, knowing three points will determine a parabola. The values of a, b, and c can be found by using three ordered pairs and then solving a system of equations.

Examples:
1. An object is thrown into the air. Suppose that initially, the object has a height of 1 foot, after two seconds it has reached a height of 5 feet, and after 3 seconds it is 6 feet off the ground. Clearly, this object was not thrown very

hard. Also, it conveniently yielded only integer values. To find the equation for this parabola, the points (0,1), (2,5) and (3,6) can be used and substituted, one at a time, for x and y in the equation $y = ax^2 + bx + c$. This gives a system of three equations:

$$5 = 4a + 2b + c \qquad 6 = 9a + 3b + c \qquad 1 = 0a + 0b + c$$

This system of three equations could be solved, giving the values of a = $-\frac{1}{3}$, b = $\frac{8}{3}$, c = 1. So, the equation of the parabola is $y = -\frac{1}{3}x^2 + \frac{8}{3}x + 1$.

2. The Gateway Arch in St. Louis is very close to, but not exactly, a parabola. It is 630 feet across at its base and 630 feet tall.[34] If (0,0) is taken to be the location of one base, two other points on the arch would be (630,0) and (315,630). These points can be substituted into the form $y = ax^2 + bx + c$ to generate a system of three equations.

$$0 = a \cdot 0^2 + b \cdot 0 + c \qquad 0 = a \cdot 630^2 + b \cdot 630 + c$$
$$630 = a \cdot 315^2 + b \cdot 315 + c$$

Solving this system gives values of a = –0.006349206, b = 4, and c = 0. So, a good approximation for an equation of the Gateway Arch is $y = -0.006349206x^2 + 4x$.

3. There is an interesting theorem in geometry which states that for any three non-collinear points, there is a unique circle that passes through them. Given three points, how could the equation of this circle be found? To find it, use the general form of a circle, $x^2 + y^2 + ax + by + c = 0$. There are at least three ways to find the equation, and for any of them, one would need to solve a system of equations. The three each have about the same level of difficulty—not too difficult, but each a bit cumbersome.

Method 1—As shown above, the general form for a circle, $x^2 + y^2 + ax + by + c = 0$ could be used. Then each of the three points is substituted into the general form. Three equations are obtained and can be solved to find a, b, and c.

Method 2—This method is a good review of a number of mathematical concepts. Given three points, take a pair of the points and find the equation representing the perpendicular bisector of the segment joining them. This will involve using the midpoint formula, slope formula, and point-slope formula for a line. Then do this again with a second pair of points (one point will be used twice). This will give two linear equations that can be solved as a system. The solution represents the center of that circle. Finally, use the distance formula to find the distance between the center and any of the three original points to find the radius of the circle.

Method 3—The distance from the center to each of the points must be the same. Therefore, set up the distance formula three times with each of those points and the missing center. That will give three expressions that are all equal. Twice, set any two expressions equal to each other and solve that system to find the center. Then, once again, use the distance formula to find the radius.

4. Planetary orbits are elliptical. The general form of an ellipse is $ax^2 + by^2 + cx + dy + e = 0$. Finding the equation of the ellipse using this method would call for finding five ordered pairs in the path of the planet and then solving a system of five equations.

Money Problems

Money problems are often found in situations that yield systems of equations.

Examples:

1. A sum of three thousand dollars was invested in two different accounts. One made 4 percent and the other made 6.5 percent in interest for the year. The year-end statement says a total of $150 interest was made. How much was invested in each account?

The amounts, x and y, in each account could be found with these equations:

$$x + y = 3,000$$
$$0.04x + 0.065y = 150 \text{ [Answer: } x = \$1,800 \text{ and } y = \$1,200]$$

2. Working the ticket booth at the school play, you collect money and hand the people their tickets. The cost is $5 for adults and $3 for students. Unfortunately, you didn't keep track of how many you sold of each. However, looking at what is left of your roll of tickets, you know that you sold a total of 191 tickets. From the amount in your cash box, you know you made $815. How many adult tickets, x, and student tickets, y, were sold?

$$x + y = 191$$
$$5x + 3y = 815 \text{ [Answer: } x = 121 \text{ and } y = 70]$$

Distance, Rate, and Time

Distance/rate/time problems are not popular. That is mostly because they have been the source of countless ill-conceived story problems. What is the student to think when he reads, "David rode his bicycle 3 miles at a speed of 10 miles per hour. Steven left 10 minutes later than David, but was riding at 12 miles per hour. When will Steven catch up to David?" The student in the math class probably doesn't care. David and Steven probably don't even care. If they did, they probably wouldn't try to figure out the math while riding their bikes. However, problems can be designed that might have more relevance.

Example:

Any frequent flyer knows that there can be quite a difference in flying times depending on whether the plane is flying with or into the wind. For long flights, there can be an hour or more difference. Suppose you are flying round trip, with each direction being 3,000 miles. One direction took 4 hours and one took 5 hours. Assuming this difference in time is due solely to the wind, what was the speed of the plane and the speed of the wind?

If x is what the plane's speed would be with no wind and y is the speed of the wind, the faster flight is the flight with the wind—a speed of x + y. The slower flight was into the wind—a speed of x – y. Since Distance = Rate · Time, we can obtain two equations.

$$3,000 = (x + y) \cdot 4$$
$$3,000 = (x - y) \cdot 5 \text{ [Answer: } x = 675 \text{ mph and } y = 75 \text{ mph]}$$

Taxi Cab Rates

Taxi cab companies often have a set charge and then a certain amount per mile. The following example uses actual figures from two services.

Example:

Taxi Service A charges a flat fee of $2 plus $2.30 per mile. Taxi Service B charges $2.25 and then $1.80 per mile. Is there a mileage at which the charges would be the same?

Two equations that represent the companies are:

Taxi Service A: $y = 2.3x + 2$
Taxi Service B: $y = 1.8x + 2.25$

The substitution method would probably be the easiest way to solve this system. However, since they are both in slope-intercept form, it might be interesting to solve them using the graphing method. Either way, it turns out the magic distance is one-half mile. Students could substitute 0.5 and other values to verify the validity of this answer. It could also be asked, for what distances does Taxi Service A cost more than Taxi Service B? This would be found with the inequality $2.3x + 2 > 1.8x + 2.25$.

Similar problems can be found in other areas. The rates for rental cars may have a flat rate along with a rate per mile or rate per day. It can be determined at how many miles or days it would be less costly to rent one car than the other.

Many workers, such as car salespersons and real estate agents, might work on a partial commission basis. A real estate agent might receive a base salary of $30,000 per year in addition to 5 percent of all sales. How might that compare with an agency that offers $25,000 and 6 percent on sales?

Break-Even Point

A break-even point for a company is the point where income and expenses are equal. Anything beyond the break-even point would be profit. Costs can be classified as fixed or variable. Fixed costs are set amounts and probably include items such as rent, taxes, and employee's salaries. The number of items the company produces will vary, and thus will be a variable cost.

Example:
Your company sells games that cost $4.20 each to produce. On top of that are fixed costs of $100,000 per year. Each game sells for $5. Where is the company's break-even point?

If x is the number of games produced and y is the dollar amount, equations representing the company's cost and the company's income could be written. The equation $y = 4.2x + 100,000$ would represent the cost and $y = 5x$ would represent the income.

Solving these equations gives answers of $x = 125,000$ and $y = 625,000$. So, if 125,000 games are produced, the cost and income are both $625,000 for the year.

Again, this problem could be turned into an inequality application by asking how many games would have to be made in a year to make a profit. The inequality $5x > 100,000 + 4.2x$ would find that range of values.

Supply and Demand

The laws of supply and demand are actually two laws. The law of supply is that the supply of an item and its price have a direct relationship. If the price of an item goes up, companies will make more of that item. The law of demand has an inverse relationship. A higher price for an item will reduce its demand. A lower price would increase it. "It is possible for supply and demand to be equal. In order for this to happen the amount of supplied products or services must equal the demand for those products and services. If this is attained, the economy is balanced in a state of equilibrium."[35]

Television Screen Dimensions

The "aspect ratio" gives the ratio of the width to the height of a television screen.

Examples:
1. For televisions, the aspect ratio is often 4:3. To have the picture on a 25 inch screen, what should be the actual width and height of the television screen?

Television screen sizes are measured by the length of the diagonal going from one corner to the opposite corner. Using the Pythagorean Theorem, along with the width (x) and height (y) of the television screen, gives the equation $x^2 + y^2 = 25^2$. Also, having an aspect ratio of 4:3 means that $\frac{4}{3} = \frac{x}{y}$. A substitution with $x = \frac{4}{3}y$ could be made into the first equation. Doing so gives $\left(\frac{4}{3}y\right)^2 + y^2 = 25^2$. This equation would yield values of x = 20 and y = 15.

2. IMAX movies typically have an aspect ratio of 1.43:1. We wish to design a 60-inch television with aspect ratio such that an IMAX movie would fill the screen without distortion. What would the width and height need to be?

This gives two simultaneous equations:

$$x^2 + y^2 = 60^2 \text{ and } \frac{1.43}{1} = \frac{x}{y} \rightarrow (1.43y)^2 + y^2 = 60^2 \rightarrow y = 34.3847$$

The television should have a height of 34.38 and a width of 49.17 inches.

Global Positioning System

The Global Positioning System makes use of orbiting satellites to find the location of any person that has a GPS receiver. It is called a receiver because it is receiving radio signals sent out by the satellites. There are twenty-four GPS satellites constantly orbiting the Earth. Although only three are necessary, there are always at least four satellites that are in contact from any point on the Earth.[36] Knowing the speed of the radio signal being sent, the time the signal takes, and using the d = rt formula, the distance from each satellite can be computed. (Because of the movement of the satellites relative to the Earth, time will appear slower, thus Einstein's theory of relativity also has to be taken into account.) The distances from each satellite describe spheres. The intersection of the spheres marks the location on the Earth.

II

GEOMETRY

Logic

From the time of Euclid and Pythagoras, geometry and logic have gone hand in hand. Logic has been used to advance geometry. But geometry can also be used to develop logic in students. In 300 BC, Euclid wrote *The Elements*. He started with twenty-three definitions, five postulates, and five "common assumptions," and from there proceeded to develop much of mathematics used for the next several hundred years.

Charles Dodgson, a.k.a. Lewis Carroll, the writer of *Alice in Wonderland*, was actually employed as a lecturer in mathematics. He lectured at Christ Church, a part of Oxford University, in England. He wrote logic puzzles for his students to analyze and to draw conclusions from. He wrote[1]:

No ducks waltz.
No officers ever decline to waltz.
All my poultry are ducks.
[Conclusion: My poultry are not officers.]

Or,

Babies are illogical.
Nobody is despised who can manage a crocodile.
Illogical persons are despised.
[Conclusion: Babies cannot manage crocodiles.]

All writers, who understand human nature, are clever.
No one is a true poet unless he can stir the hearts of men.
Shakespeare wrote "Hamlet."
No writer, who does not understand human nature, can stir the hearts of men.
None but a true poet could have written "Hamlet."
[Conclusion: Shakespeare was clever.]

To learn Geometry means to learn to think logically. Geometry, maybe more than any other topic of study, lends itself to the study of logic. While geometry proofs will never win any popularity contests among students, being able to follow and to create a logical argument is an invaluable skill. It is a skill not everyone seems to have. Just as important as recognizing logical arguments, is being able to recognize cases of illogic and what makes them so.

Inductive Reasoning Flaws

Both inductive and deductive reasoning are used in a geometry course. Both types can be flawed. Inductive reasoning goes awry when we look at patterns and assume those patterns will continue. People say that no two snowflakes are alike. There has yet to be found two snowflakes that are alike, but that doesn't make it so. People say lightning never strikes in the same place twice. It actually does. Otherwise, we might as well throw away lightning rods after one use. Patterns often do hold up. We know which diets seem to work, which foods are good for you, how to make medical prognoses, and how to predict the weather, often by the use of inductive reasoning. Researchers use inductive reasoning when conducting trials or just looking to past history in order to make predictions.

But, sometimes patterns don't hold up. Consider the population of U.S. residents that are foreign born. In the 1910 U.S. Census, 14.7 percent of the population was foreign born. There seemed to be a definite pattern developing in the following decades.

Census	1910	1920	1930	1940	1950	1960	1970
% Foreign Born	14.7%	13.2%	11.6%	8.8%	6.9%	5.5%	4.7%

A teacher could have students write a linear equation modeling these statistics to predict future decades. However, after 1970, things change.

Census	1970	1980	1990	2000	2010
% Foreign Born	4.7%	6.2%	7.9%	11.1%	12.9%

There are many other cases where inductive reasoning can lead us down the wrong path. What if a coin is flipped 88 times and it came up heads each time? We might well assume it is a two-headed coin. After all, doesn't something happening 88 times in a row constitute enough of a pattern to "prove" this assumption? A person might have started following the UCLA Bruins basketball team on January 29, 1971, at the start of what would turn out to be an 88 game winning streak. Yet, three years later, that pattern ended.

Here is another example: choosing any number of points on a circle and connecting those points with line segments. Now count the regions into which the circle is divided. With one point, there is nothing to connect it to, so there is one region. Connecting two points makes two regions. With three points we can obtain four regions. There is a pattern of 1, 2, 4, 8, 16. An astute math student might guess this pattern is based on powers of two and assume the next number will be 32. The pattern breaks down however, as it continues with a sequence of 1, 2, 4, 8, 16, 31, 57, 99,…. Things were going so well, and so predictably, for the first five terms, then it fell apart. If n is the number of points, the relationship looked like the pattern was going to be based on $f(n) = 2^{n-1}$. It isn't. It turns out there actually is a pattern. It is the far less obvious $f(n) = 1/24(n^4 - 6n^3 + 23n^2 - 18n + 24)$.[2]

Ponzi schemes sucker people in by what looks like good inductive reasoning. Give someone, say his name is Bernie, your money. He claims you will get it back and then some. How can he pull that off? As more and more money is collected, he gives it to the first investors. They are making a nice profit and they tell people how great Bernie and his investments are. People see the pattern that people investing with Bernie get rich. They want to invest, too. He can keep this going for a while, but you find out that at some point Bernie made off with your money and only those initial investors made anything at all.

Deductive Reasoning Flaws

Deductive reasoning begins with postulates and logically proceeds to prove other statements. Thomas Jefferson used this concept when he started with truths that were self-evident and then proceeded to state why this meant a need for independence from England. Lawyers make a living from their skill at using deductive reasoning.

What is the best kind of reasoning? Since deductive reasoning is the basis of geometric proofs, students start feeling like deductive reasoning must be the correct answer. However, to take a set of premises and base a deduction from them is only as sound as those original premises. As computer programmers say, "Garbage in, garbage out." Since many geometry students have written "proofs" that have not turned out to be true, deductive reasoning must not be flawless. In fact, a good starting point to studying deductive flaws is with their own proofs. There are plenty of other examples.

The food pyramid for what makes for a healthy diet has changed over the years. The original pyramid was flawed due to bad information given to the USDA by lobbyists representing various food organizations.[3]

In the past, funding of medical research discoveries have led to additional funding for the research team only to find out that those results had been doctored. (Sorry.)

Isaac Newton said that force is equal to mass times acceleration. Adding more force will cause a proportional increase in acceleration. Albert Einstein changed how we look at the universe by postulating that the speed of light is a constant value regardless of the frame of reference.

Probably the most important example of deductive reasoning issues was committed by the king of logic, the mathematician Euclid.

Non-Euclidean Geometry

Mathematics use axioms or postulates that are accepted because they are just obviously true. Euclid began his writings in his book, *The Elements*, with nine postulates. Who can argue with Euclid's fourth postulate—"All right angles are congruent"? The fifth postulate seemed fine as well for quite a long time. "That, if a straight line falling on two straight lines makes the interior angles on the same side less than two right angles, the two straight lines, if produced indefinitely, meet on that side on which are the angles less than the two right angles."[4] Granted, a little wordier, but still taken as being true for centuries. And it is absolutely true, as long as we are making our drawings on a flat surface. Called the Parallel Postulate, it doesn't hold up for what we now call non-Euclidean geometry.

The geometry taught in high school classes is primarily, if not entirely, Euclidean. The surface of a sphere, such as the Earth, is non-Euclidean. There are many things that happen with lines drawn on the Earth's surface that could not happen in a Euclidean space. Lines of longitude, seemingly parallel when crossing the equator, do not only intersect, but do so twice (at the North and South Poles). Triangles add up to more than 180 degrees. (The triangle bounded by lines of longitude at 0 and 20 degrees and the Equator has angles of 90 + 90 + 20 = 200 degrees.) Many other things take place that differ from what we are used to.

This has caused a lot of issues in the past. The size of Greenland isn't really even close to the size of South America. However, you wouldn't know it by looking at most maps. Why don't they just draw it right? Because they can't. Trying to duplicate the non-Euclidean surface which is Earth onto a flat map can't be done without distortion. If you tried to peel the surface map of a globe and lay it flat on a table, you would have trouble. Notice after peeling an orange and laying the peel on a table, it won't lie flat.

And Earth isn't the only problem. The case with spheres is called Riemannian geometry. Another non-Euclidean geometry is called hyperbolic geometry and could be likened to drawing on a saddle. Albert Einstein showed that space itself is curved and obeys rules governing non-Euclidean geometry. Gravity isn't the nebulous force that Newton thought it was, but a warping of the geometry of space so that it obeyed the rules of non-Euclidean Geometry. What seemed to be a self-evident axiom turned out to be anything but.

Boolean Algebra

The previous are examples of logic gone wrong, but there are plenty of examples of logic going right. We employ logic, to some degree, every day. However, in some fields, logic overflows. In the mid-1800s George Boole developed a logical system that has had great application to the modern world. For most, "algebra" conjures up thoughts of equations, numbers, variables, and arithmetic operations. Boolean algebra deals with "and," "or," "not," "if," and makes decisions as to the truth or falseness of given statements. It was almost a hundred years after the development of Boolean algebra, also known as symbolic logic, that it became of much real use, and then it became invaluable. Consider the statement "1 + 1 = 2 and dogs can fly." This is a false statement. On the other hand, "1 + 1 = 2 *or* dogs can fly" is a true statement. Neither statement seems very history making. However, those statements are similar to electric circuits with light bulbs A and B wired in. If wired in parallel, if there is a break in bulb A *or* bulb B, the circuit will still be complete. On the other hand, a break in bulb A *and* bulb B will cause the circuit to break and no electricity will be able to flow through.

Telephone switching makes connections between telecommunication devices and allows them to communicate. In the past, this was physically done by telephone operators. In 1938, it was found that Boolean algebra could be applied to the design of circuits involving switches. The true-false model was used with truth correlating to a connection and false correlating to no connection.

Computer Programming

To do much in the way of computer programming could be quite time consuming, but the need for logic can be seen in simply having the ability to interpret simple computer programs. Students should be able to see the following as a series of logical steps not unlike a geometric proof. With a just a little help, they should be able to determine the output of this and similar programs.

```
Step 1: Let n = 1
Step 2: Print n
Step 3: Let n = n + 1
Step 4: If n < 100 then go to step 2
Step 5: End
```

There is a very strong connection between logic and computer programs themselves (called the Curry Howard isomorphism) that basically states that a correct computer program is equivalent to a proof of a logical formula.[5]

Ratios and Proportions

Batting Average

A batting average is simply a ratio of a player's hits to his times at bat—typically written as a decimal rounded off to thousandths place.

Examples:

1. If a player gets 15 hits in 57 at bats, what is his batting average?

$x = \frac{15}{57}$ [Answer: .263]

2. How many times was a player up to bat that had a batting average of .311 and 159 hits?

$\frac{311}{1000} = \frac{159}{x}$ Answer: 511 at bats]

Inflation

Inflation is the rate that prices would increase during a year's time.

Examples:

1. If inflation is at 2.3 percent, how much would you expect the price of a $217 cell phone to increase in a year's time?

$\frac{2.3}{100} = \frac{x}{217}$ [Answer: $4.99]

2. What is the inflation rate if items costing $23 have a price of $24.50 the following year?

$\frac{24.5-23}{23} = \frac{x}{100}$ [Answer: 6.5 percent]

What are the most watched movies of all-time? *Star Wars Episode VII* leads the way if you go strictly by the dollars it has made. That hardly seems fair, though, when considering that movie tickets used to cost a quarter. A much fairer way is to account for the average movie price for a given year. Sites exist that give the price of the average movie ticket for any given year. A possible project would be to use proportions to calculate the top movies of all-time in today's numbers.

The very first *Star Wars* movie made $461 million dollars. However, that was far back in 1977 when the average ticket was $2.23. How much would it make today assuming ticket prices of $9?

Inflation is the increase in prices per year. In fact, any measurement containing the word "per" can be written as a ratio of two amounts. Many examples of these exist: feet per second, miles per gallon, dollars per gallon, miles per hour, yards per carry, etc.

Examples:

1. At 63 miles per hour, how far would you travel in two and a half hours?

$\frac{63}{1} = \frac{x}{2.5}$ [Answer: 157.5 miles]

2. If gasoline is $4.26 per gallon, how many gallons can be bought with $20?

$\frac{4.26}{1} = \frac{20}{x}$ [Answer: 4.7 gallons]

Animal Population Estimates

Animal populations can be estimated by a mark-recapture method. Animals can be caught and marked in some way, with a leg band for example, and released back into the wild. Later, another group of animals is caught. By determining the ratio of banded to unbanded animals in that second catch, an estimate of the total number of animals could be determined.

Example:

Twenty red foxes are caught, banded, and then released. Two weeks later, 53 red foxes are captured. Only 7 of those have bands. The proportion $\frac{20}{x} = \frac{7}{53}$ could be used to estimate the total number of red foxes living in that area.

Currency Exchange Rates

Each country has a particular currency used to conduct business. Money can be exchanged based upon the current exchange rate. This could be done a variety of ways, but is usually done at banks.

Example:

You come back from a trip to England with 257 British pounds and would like to estimate how many dollars that might be after you exchange them. If the current exchange rate is that each pound is worth 1.6168 dollars, the amount could be found by using the proportion: $\frac{257}{1} = \frac{x}{1.6168}$.

Distance to the Moon

Some ingenious geometric methods have been used in the past to determine the distance to the Moon. The Greek Aristarchus came up with a method 2,000 years ago. Although he came amazingly close to the actual figure, more exact measurements have been possible in the past century with the invention and use of radar and, a bit later, of lasers. We know the speed of light is approximately 186,000 miles per second. If a radar pulse or a laser could be bounced off the Moon, the trip could be timed and a proportion set up to calculate the distance. During the first mission to the Moon, our astronauts set up a reflector on the Moon's surface so this could, in fact, be done. It turns out it takes about 2.5 seconds for the round trip. Using $\frac{186,000}{1} = \frac{x}{2.5}$, and then dividing by two, to find the length of a one-way trip, the distance to the Moon could be determined. Using a greater number of significant digits, scientists have found the distance to within a single millimeter.

Musical Scales

In the fifth century AD, Pythagoras developed a theory of how musical scales worked. He found that a plucked string made a musical tone of a certain pitch. If he plucked a string half that length it produced a pitch that was an octave higher. Strings two, three, or four times as long produced other tones an octave apart. Other tones in a musical scale corresponded to other lengths. Although he didn't know it at the time, there was a relationship between the plucking of a string and the number of cycles per second of sound waves that were produced due to the vibration of the string.

Pythagoras' ratios corresponded to notes of scales. He developed ratios between the notes based on the lengths of the strings producing the notes. For example, from C to the C an octave above was a ratio of 2:1. There were ratios for each of the intervals in between. From C to D was the ratio 9:8, C to E was 81:64, C to F was 4:3, and so on until back to C, a full octave above and the ratio 2:1.[6] There has been some tinkering done with those relationships in the centuries since, mainly by making slight changes in some of the ratios. Primarily in use today, equal tempered scales are not based on different ratios, but on a division of the frequencies of the twelve notes[7] based on the function $f(x) = 2^{\frac{x}{12}}$. (More on this relationship in another section of this book.) This relationship yields results very close to, although not identical to, Pythagoras' ratios.

Recipes

Cooking recipes could be made to feed more, or fewer, people by changing the amounts of the ingredients.

Example:
A recipe calls for 8 ounces of uncooked pasta and 6 garlic cloves. If you want 1½ times the recipe, you could use the following:

$$\frac{3}{2} = \frac{x}{8} \quad \text{and} \quad \frac{3}{2} = \frac{x}{6}$$

Gas Laws

There are a number of laws relating to gasses that show a proportional relationship. One of these is Charles' Law. Jacques Charles was the uncommon combination scientist/balloonist. He co-developed, and rode in, the very first hydrogen balloon in 1783. Manned balloon flight was at its golden age and Charles played an important part. His law states that temperature, in degrees Kelvin, and volume of a gas are proportional.[8]

Joseph Louis Gay-Lussac (pronounced "LUH-sak") was a French scientist born in 1778. It was he who found that water is two parts hydrogen to one part oxygen. Additionally, he found a law, Gay-Lussac's Law, which states that the temperature, in Kelvin, and the pressure of a gas, varies directly.[9] Because of Gay-Lussac's Law, it is said that tire pressure should be measured before driving, as the act of driving will cause the air in the tire to heat up and thus have increased pressure. It is also why the labels on aerosol cans will often state, "Do not incinerate." The greatly increased heat leads to greatly increased pressure, possibly causing an explosion.

You might be aware that a basketball left in the cold doesn't bounce very well. The decreased temperature causes the air in the ball to shrink a bit, decreasing the air pressure inside the ball. If it is brought inside or run under hot water, it will regain its volume and its bounciness.

Examples:

1. A basketball was left outside at 8 degrees Celsius. Its volume was 444 cubic inches. When it was brought inside, it warmed up to 32 degrees. What is its volume now?

First, use the formula K = C + 273 to convert from degrees Celsius to degrees Kelvin. Then, using Charles' Law,

$$\frac{444}{281} = \frac{x}{305} \quad \longrightarrow \quad x = 482 \text{ cubic inches}$$

2. NFL Rule 2.1 states in part that "The ball shall be made up of an inflated (12½ to 13½ pounds) urethane bladder."[10] It is claimed that the Patriots football team has underinflated the ball to try to gain an advantage. A league report claims that at least one of the balls was inflated to only 10.5 pounds per square inch. Suppose the Patriots then claim that according to Gay-Lussac's Law the football would naturally lose pressure as it was brought from the warmth of the locker room to the field. If the locker room was 75° F (297.04° K) and the field temperature was 40° F (277.59° K), might the Patriots' claim be true that the football was actually within the guidelines of 12.5 to 13.5 pounds per square inch at one point? [Answer: 11.2 psi, so no.]

Bicycle Gear Ratios

A bicycle might have only one gear ratio, or it could have ten, or twenty-four. A gear ratio of 1:1 means that for every pedal stroke the rider makes, the rear wheel makes one revolution. If the ratio is 0.6, then every pedal stroke means the rear wheel makes six-tenths of a revolution. (The same is true for the front wheel, but the rear wheel is the driving force. It can be seen that the gears are connected to the rear wheel and the front wheel is literally just going along for the ride.) A low ratio like 0.6 is good for accelerating and climbing hills. A low gear is not so good if the rider is on flat terrain and wants to go fast. It would be possible in a low gear, but there would have to be some very fast pedaling. If the rider wanted to go fast, a ratio of 3:1 might be more helpful. Each pedal stroke moves the tire through three revolutions. Good for speed, but would be tough to maintain going up a steep hill.

So where do those ratios come from? There are two chain wheels—one in the front and one in the back. Taking the case of a ten speed bike, there are two chain wheels in the front and five in the back. To find the gear ratio the bike is in, look at the chain wheels that the chain is riding on and count the number of teeth. The ratio is the number of teeth on the front chain wheel to the number of teeth on the back chain wheel.[11] Suppose there are 36 on the front and 16 on the back. The ratio is 36:16 or 2.25. Each pedal stroke causes the bike tire to make 2.25 spins.

Similar Figures

Map Scales

Maps that are drawn "to scale" have a scale located somewhere on the map that allows a person to find straight line distances.

Examples:

1. Suppose a map states that 2" represents 105 miles. What is the distance from Portland to San Francisco? Using a ruler, say the map distance between the cities is 9.75 inches. That makes the distance between the cities:

$$\frac{2}{105} = \frac{9.75}{x} \quad \rightarrow \quad x = 511.9 \text{ miles.}$$

2. A map of the United States is to be drawn on a piece of paper that is 2 feet wide. A scale needs to be chosen so it will fit the paper. Figuring it is 2,500 miles across the U.S., what should 1 inch equal in miles to be sure of a good scale in terms of the width? [Answer: At least 104.1 miles]

The scales listed on maps can similarly be used in many other situations. Dictionaries might have a picture with a scale at the bottom. There might be a picture of an elephant and below it the caption states it is in 1:32 scale. With that scale and measurements taken of the picture, real-life values of the elephant's height, width, leg length, etc. could be found.

A model kit of a jet plane might say it is a "1:72 scale." Again, by taking measurements of the model, and by using proportions, the size of the actual plane could be found.

Aspect Ratio

A television's aspect ratio is the ratio of the width to the height of is display. Older style televisions had an aspect ratio of 4:3. The newer models have a ratio of 16:9. Since those ratios are not the same, the television screens are not similar rectangles. One can clearly see the difference by looking at an older television screen which is nearly a square shape, modern televisions, and movie theater screens which are much wider than they are tall. In the 1950s, Hollywood came out with Cinemascope to compete with television.

Those movies had a ratio of 2.35:1. Another innovation was VistaVision which had a ratio of 1.85:1. There are many others types of screens: IMAX screens, U.S. movie screens, European movie screens, computer screens, iPhone screens, and photographic film—all different. Thus, the picture shown on your screen will suffer some distortion, or there will have to be some other method to get the picture and screen to preserve shapes.

So what can be done when a movie made for a theater screen is shown on television? There have been various ways of dealing with this issue. One way of dealing with the problem is to give in to some distortion. Thus the phrase, "This movie has been modified from its original version." Since 4:3 is not the same ratio as 16:9, to force the picture into a different ratio will cause the picture to look not quite right. Another method is called Pan and Scan. Decisions are made about where the main part of the action is, making sure that it is kept, while leaving out the rest. Again, this has its drawbacks. The same idea, but easier for the movie makers, is just to lop off the outer ends of the scenes. When viewers see two people talking in a scene and they are both barely visible at the far ends of the screen, this is likely what happened. You would have seen them if your television had the same aspect ratio as the original film. Letterboxing displays the entire scene. (Letterboxing is so named because the ratio is approximately that of a typical envelope.) It keeps the aspect ratio intact, but does so by inserting those annoying large black bars displayed at the edges of the screen.

All of these problems and possible fixes come down to the issue of having to deal with non-similar figures.

Pantographs

Pantographs allow an individual to make two drawings that are the same shape, but different sizes. Or, an already drawn item can be traced and a larger or smaller image of that tracing will be made simultaneously. A pantograph accomplishes this because, as it is adjusted, corresponding angles remain congruent. Because of this, there are similar triangles within the pantograph, making corresponding parts of the drawings proportional.

Projective Geometry

Projective geometry has applications to many areas. These often, but not always, result in similar figures. A movie projector sends rays of light through a relatively tiny movie cell. Many feet away, on a movie screen, it is a geometrically similar shape, many times larger in size than the original cell.

The Earth is a sphere. Those points can be projected onto a flat surface. Imagine a light shining from the middle of a transparent globe. If it shines on a large piece of paper that is shaped as a cylinder surrounding the globe, there is a projection from the globe to the paper. Unfold that paper to have a map of the Earth. There are many styles of maps, but all are some sort of projection similar to this.

During the Renaissance, a major advance in painting was the use of perspective. Imagine lines drawn from a painter's eye to points in a scene. If a transparent canvas is put between his eye and the scene and is painted on, there is a two-dimensional representation of the three-dimensional scene. This process of projective geometry gives the realism the painter desires.

A location called a vanishing point may be used within a painting in which lines would be receding to a fixed location. Railroad tracks are parallel, but if they are drawn as parallel lines in a painting, there is no sense of them receding into the distance. There can also be a two-point perspective. A drawing of tiles laid out on a floor might be done with two vanishing points.

Road Gradients

At the top of a hill, it is common to see a sign that might say, "Caution: 6 percent grade for next three miles." This sign is posted primarily for truckers. For a car it is no big deal negotiating this descent. For a logging or coal truck, with its much greater weight and more momentum, it's a little trickier.

Examples:

1. A road has a 6 percent downgrade. What is the horizontal distance for a car to drop 450 feet in elevation?

A downgrade of 6 percent means that the road will drop 6 feet for every 100 feet of horizontal distance. So, this 6 percent grade can be thought of as a right triangle with legs of 6 and 100. A missing side of a similar triangle with one side of 450 feet can be found using a proportion.

$$\frac{6}{100} = \frac{450}{x} \quad \rightarrow \quad x = 7,500 \text{ feet}$$

2. For a road with an 8 percent downgrade, how far would a car drop in elevation after driving one mile?

One way of finding this distance is to compare two similar right triangles. One has legs of 8 and 100. The Pythagorean Theorem could be used to find the hypotenuse of this triangle. This distance turns out to be $\sqrt{10064} = 100.32$. The second triangle is one which has a hypotenuse of one mile (5,280 feet). Now a proportion can be set up comparing the drop in height to the hypotenuse of the two triangles.

$$\frac{x}{5,280} = \frac{8}{100.32} \quad \rightarrow \quad x = 421.1 \text{ feet}$$

The Pyramid at Giza

The Great Pyramid at Giza (pronounced "Geeza") was one of the seven wonders of the ancient world and the only one still in existence today. The height of the pyramid was found by Tales. Tales (pronounced "Tayleez") was a Greek philosopher/mathematician who lived six hundred years before Christ.

He was once challenged to find the height of the Great Pyramid. He did so by the fairly simple method of putting a stick in the ground, then measuring it and its shadow. He measured the length of the pyramid's shadow at the same time. Both triangles formed are similar since they are both right triangles, and being at the same time of day, the angles formed by the Sun would be the same. Accounts differ as to the exact details of this. One account states that he ingeniously waited until the stick and its shadow were the same length. He then knew that the pyramid's shadow would also be the pyramid's height.

Throw Ratio

A home theater system uses a projector to display images on the wall of a room. How far away should the viewer sit to get the full effect? If too close, the picture will appear pixilated. If too far away, the theater effect will be lost. This is known as the projector's throw ratio, v:w, where v is the distance an individual is from the screen and w is the width of the screen.[12]

Example:

Suppose a projector has this line in the directions—"If you have an LCD projector with more than 600 vertical lines of resolution, we suggest a throw ratio between 1.8:1 and 3:1." This projector is mounted in the ceiling and displays an image that is 5.5 feet across. Where should a person sit to get the full theater effect? The following proportions would give the range of distances in terms of how many feet to sit from the screen.

$$\frac{1.8}{1} = \frac{x}{5.5} \quad \text{and} \quad \frac{3}{1} = \frac{x}{5.5} \qquad \text{[Answer: Between 9.9 and 16.5 feet away]}$$

Scale Drawings

The blueprint of a building is a scale drawing of the building's layout. Although originally printed on blue paper, the term now often applies to scale drawings on any colored paper. It is a good learning activity to examine examples of scale drawings. An even better activity is to have students make their own scale drawings. A number of skills would be involved—solving proportions, finding the size of angles, and using protractors and rulers. Following are some items that could be used.

From 1861 to 1928, United States bills had dimensions of 7.42 by 3.13 inches. Bills now are 6.14 by 2.61 inches.

Not all flags were created equal. British and associated countries (Canada, Australia, Ireland, etc.) all have flags with length to height ratios of 2:1. French flags are 3:2, and German flags are 5:3.

The United States' flag has a length to height ratio of 19:10. The other ratios are also made in comparison to the height of the flag. The ratio of each stripe height to the height of the flag is 1:13. The blue portion's height compared to the flag height has a ratio 7:13 and the length of the blue portion to the flag height is 19:25.

Stop signs are regular octagons with interior angles of 135°.

The Pentagon in Washington, D.C., is a regular pentagon with a perimeter of 921 feet. The interior courtyard is also a pentagon and has a perimeter of 250 feet.

The Washington Monument has a total height of 555 feet, 5 inches. It has its 500-foot-tall bottom section and is topped with a pyramid with a height of 55 feet, 5 inches. The pyramid has a square base with sides of 34 feet, 5.5 inches. The base of the monument is a square with sides of 55 feet, 1.5 inches.

Photo Enlargements

Photographs can be enlarged to any imaginable size. Many involve some distortion. Comparing the width to length ratio of the most common sizes of photographs shows this to be the case. The common enlargement sizes of 4x6, 5x7, 8x10, and 11x14 are relatively close, but do not have exactly the same ratios. There are programs that can be used with an option to "keep the same aspect ratio," making the enlargement geometrically similar to the original. Proportions can be used to find the dimensions that would make for similar photographs.

Example:
If a 5x7 photograph is to be enlarged to have a width of 2 feet, what should its height be?
[Answer: A height of 2.8 feet, or 2 feet, 9.6 inches]

Copy Machines

For the most part, the role of a copy machine is to make replicas of an image. However, most machines have an enlargements/reductions feature which, in essence, creates similar figures. The control panel will allow the user to pick an image that is a percentage of the original. To set an enlargement of 110 percent would be to create similar shapes with a ratio between them of 11:10.

The Golden Ratio

The golden ratio and the golden rectangle are fascinating topics that could be explored in a number of ways. The golden ratio has the value $\frac{1+\sqrt{5}}{2} = 1.61803$. The golden rectangle is any rectangle which has a length to width ratio equal to the golden ratio. There are a number of applications, including both natural and man-made objects. The golden ratio is also evident in paintings, crystals, the pyramids, the human body, and in the spiral shapes of shells and galaxies.

The Fibonacci sequence makes its appearance in a number of real world areas. As one proceeds farther out with this sequence, the ratio of any two consecutive terms approaches the golden ratio.

Although we don't know the original architects' building plans, it would seem that the design of the Parthenon in Greece was purposely designed to be based on the golden rectangle. This is true of the front facade as a whole, as well as smaller portions of the building.

Angles

Diamond's Angle Cuts

The angle at which a diamond is cut makes all the difference in how entering light is reflected. A cut too shallow or too steep allows light to escape through the sides or the bottom rather than being reflected back. In 1919, Marcel Tolkowsky determined that what is known as the pavilion main angle should be cut at 40.75 degrees.[13] Other angles needed an equal amount of precision. The angles have been debated and altered in the years since, but clearly one has to know their angles to be a diamond cutter.

Fowler's Position

Fowler's position is used in the medical field to denote the position of a patient when the head of his bed is elevated.[14] Fowler's position could be 30°, 45°, 50°, etc. It is often used in hospitals to aid in patient breathing.

Earth Measure

Angle measurement is a vital part of how the world is measured. From the Prime Meridian, "lines" of longitude are numbered from 0 to 180 degrees west and 0 to 180 degrees east. Lines of latitude begin at the Equator at zero degrees and extend 90 degrees north to the North Pole and 90 degrees south to the South Pole. These measurements are based on the geometric definition that the measure of an arc of a circle is equal to the measure of the central angle that subtends it. The North Pole is at latitude 90° because the angle with its vertex at the center of the Earth and sides intersecting the Equator and the North Pole point form a 90° angle.

One degree might seem like a very small amount when drawing angles with a protractor on a piece of paper. However, if the surface used is Earth instead of a sheet of paper, one degree can encompass quite a distance. The Earth has a circumference of 24,900 miles. Thus, one degree covers $\frac{1}{360}$ · 24,900 = 69.2 miles. That is not nearly accurate enough in most cases. Degrees can be divided into 60 minutes and each minute into 60 seconds. One second on the surface of the Earth is $\frac{1}{360}$ · 24,900 · $\frac{1}{60}$ · $\frac{1}{60}$ = 0.0192 miles = 101.4 feet. Locations are sometimes given to the nearest tenth or hundredth of a second, increasing the precision. A measurement to the nearest hundredth of a second would be to place an object with an accuracy of approximately one foot.

For cities, the nearest minute would probably be fine. London, England is located at 51°31' N and 0°5' W. On the other hand, giving the location of the Brooklyn Bridge would require more precision. The Manhattan side of the bridge is over water at 40°42'28.9" N, 73°59'57.2" W. It encounters land again at coordinates 40°42'15.3" N, 73°59'41.0" W.

Latitude and longitude are sometimes listed in decimal form and sometimes broken apart into degrees, minutes, and seconds. Most calculators have a way to make that conversion, or it can be done with some basic arithmetic.

Examples:
1. What is 40°42'28.9" in decimal form?
To convert seconds to minutes, 28.9 ÷ 60 = 0.48167. The total number of minutes, then, is 42.48167. To convert those minutes to degrees, 42.48167 ÷ 60 = 0.70803. So, the decimal form is 40.70803 degrees.

2. The Alamo is at latitude 29.425°N. What is its location to the nearest minute?
0.425° · 60 = 25.5 minutes and 0.5' · 60 = 30 seconds. The answer is 29°25'30"

Students will notice that although angle and time measure are generally unrelated, they have some similarities. Both employ minutes and seconds as units of measure and conversions are based on the number sixty. In the past, other cultures have often used a base sixty system. While most of us prefer a decimal notation over fractions, decimals are the relative newcomers. Fractions have been used for thousands of years while decimals came into prominence in the past 500 years. If one is in the position of having to use fractions, it is helpful to have a denominator that is divisible by as many numbers as possible. In the past, sixty was considered a pretty good choice, having factors of 1, 2, 3, 4, 5, 6, 10, 12, 15, 20, 30, and 60.

Distance Running

This is a bit of a departure from angle measure, but mathematically, very similar. Finding the average pace of a distance runner is roughly the same method used to convert straight degrees to degrees, minutes, and seconds. If there were 100 seconds in a minute, a 5:23 mile would also be a 5.23 mile. Alas, it is a bit more complicated than that.

Example:

A distance runner is in a 3.1 mile race and has a time of 21:17. What is her pace per mile?

First, the time needs to be changed to a decimal form for minutes: 21:17 = $21\frac{17}{60}$ = 21.28333 minutes. To find the pace per mile, 21.283 ÷ 3.1 = 6.8656 minutes per mile. In the more familiar minutes and seconds, 0.8656(60) = 52 seconds. So, her running pace is 6:52 minutes per mile.

Navigating by the Stars

Students think of protractors as the lone tool used to measure angles. Actually, in the past, a number of devices have been used.

Transits and theodolites are similar devices that have been, and still are, used by surveyors. They have a telescope attached and can measure horizontal and vertical angles. Theodolites are generally more precise, having an accuracy up to one-tenth of a second of an angle.

Quadrant, sextants, and octants have been used by navigators for centuries and are still valuable backups in cases where GPS signals or other electronic devices fail. The quadrant is so named because the shape is one quarter of a circle. Coming a century or so later, the 1700s saw the development of the octant (one-eighth of a circle) and the sextant (one-sixth of a circle).

In the Northern Hemisphere, the guiding light is Polaris, also known as the North Star. The Earth is constantly rotating on its axis. As a result, all the stars appear to move, except one—the one directly over the North Pole—Polaris. If a person standing at the North Pole looked straight up, there would be Polaris right overhead. In fact, you would have to look up ninety degrees. Imagine you are standing on the Equator. You wouldn't have to look up at all. In other words, you would look up zero degrees. Similarly, a person who has to look up forty-two degrees to see Polaris is on the 42°N line of latitude. That fact, along with the devices mentioned above, was invaluable to sailors on an ocean that had no other landmarks to guide them.

Finding the line of longitude for your ship was a little trickier. However, sailors did know that, as you traveled east or west, events such as sunrise, sunset, and noontime were all shifted in time. The Earth rotates 360 degrees in a day and 15 degrees (360 ÷ 24 = 15) in an hour. By taking measurements, navigators could determine their longitudinal position. Suppose you set your clock to Greenwich, England, time and sailed west. You note that sunset is at 8 p.m., according to your clock. However, for that date, you know that the Sun would normally set at 6 p.m. back at Greenwich. You are two hours ahead and must be at 30 degrees longitude west of Greenwich.

The drawback to all this was the unreliability of timepieces. Changes in temperature, pressure, and the rocking of the ship all rendered the clocks then in use, ineffective. As Dara Sobel points out in her book *Longitude*, this set off a major endeavor (there was prize money involved) to develop a timepiece that would be reliable. In the late 1700s, John Harrison found the solution and navigation was changed forever.

Parallel Lines

Scissor Lifts

A scissor lift is a hydraulic lift system with two intersecting supports connecting its two bases. The bases remain parallel as it rises or lowers. The supports form two triangles between the bases. These supports stay the same length as the lift goes up and down and, as a result, stay in the same ratio. Because there is a pair of vertical angles, the two triangles are similar by the SAS similarity theorem. Because of this, the corresponding angles are congruent, which means the bases will remain parallel even while being raised or lowered. This is good news for anyone on the scissor lift.

Adjustable Backboards

Adjustable backboards can be raised or lowered so the rim is at various heights. However, these height adjustments need to be done so that the backboard remains parallel to the pole and the rim stays parallel to the ground. Note that the metal bars that connect the backboard to the pole stay the same length while it is being raised or lowered. The bars form a quadrilateral that, because the opposite sides are congruent, is a parallelogram. So the backboard stays parallel to the pole, making the rim stay parallel to the ground.

Tackle Boxes

Tackle boxes and other storage containers work much the same way as the adjustable backboards. When opened, trays are raised up to be displayed. The trays stay parallel as it is being raised. Typically, there is a pair of metal bars on the sides connecting the trays. Again, it can be seen that they form a quadrilateral whose opposite sides are congruent. Though this quadrilateral changes shape, it remains a parallelogram, thus the trays will be parallel.

Parallel Rulers

One more example: A parallel ruler can be used as an aide in navigation, or for general use in drafting. It is made up of two connected bars or rulers. One bar can be held secure while the other bar is moved. If a heading of NNE is desired, one bar is laid onto that compass built into the map. The other bar is moved to be aligned with your current position. The line of that bar will then also show the line that is NNE. These parallel rulers are usually made from transparent plastic, so it is easy to see the map while using this device.

These lines are parallel because, with two pairs of opposite sides being congruent, a parallelogram is again formed. This makes the opposite sides parallel, regardless what position the ruler is in.

Parking Lots

A business just repaved its parking lot. Slanted lines now need to be painted in to show the parking spaces. A tape measure is used to make sure the lines are all the same length and the same distance apart. These lines also need to be parallel, though. How does that get accomplished? One way would be to nail two boards together at some predetermined angle. This device is then used to lay out the angle for each line. Since these corresponding angles are all congruent, these lines will be parallel to each other.

Pythagorean Theorem

Changing Lanes

There are races at track meets in which runners stay in their own lane for the first curve, but after the runners come to the first straightaway, they are allowed to cut in to the first lane. Typically, runners cut in as soon as possible. But is that the best strategy? Why not gradually cut in by making for the next turn along a straight line path? The shortest distance between two points is a straight line, after all. This situation makes for a nice use of the Pythagorean Theorem. Suppose a runner is in an outer lane, four meters from the inside lane. How far is it to make it to that coveted inside lane before the next turn? Option 1—if you cut in right away, the distance is $4 + 100 = 104$ meters. Option 2—make a straight line to lane one of the next turn. That distance would be $\sqrt{4^2 + 100^2} = 100.8$ meters. This isn't a huge amount, but many races have been won or lost by a 3.2-meter difference.

Throw to Second Base

In major league baseball, the bases are 90 feet apart. To get from one base to another requires a left turn of 90 degrees. A runner on first base attempts to steal second base. How far would the catcher's throw have to be? The Pythagorean Theorem yields $\sqrt{90^2 + 90^2} = 127.3$ feet.

Speaking of diagonals and sports, a football field is 100 by 53⅓ yards. The Pythagorean Theorem would give the length of the diagonal of the field. [Answer: 113⅓ yards]

Olympic size swimming pools are 50 by 25 meters. If someone wanted to swim practice laps as far as possible without having to make a turn, the formula would tell you that as well. [Answer: 55.9 m]

Theory of Relativity

Pythagoras' theorem is not limited to two dimensions. To find the distance from a corner of a room to its opposite corner, use the formula $d = \sqrt{a^2 + b^2 + c^2}$. While there are only three spatial dimensions, Einstein extended the Pythagorean Theorem to four dimensions by considering time to be the fourth dimension.

Guy Wires

In order to support a structure, guy wires stretch from near the top of an object and are anchored to the ground. The objects needing support could be radio or television towers, wind turbines, ship masts, tents, or trees. These are all objects that would be susceptible to falling in a gust of wind.

Example:
Three guy wires are to be attached to a 100-foot tower, each anchored 80 feet from the tower's base. How much wire is necessary?

$$3\sqrt{100^2 + 80^2}$$ [Answer: 384.2 feet]

Doorway Space

You are planning to buy a very big, big-screen television. After making the purchase, you are a little concerned that it might not fit through the door. The front doorway space is 29 by 80 inches. There is actually more space than 80 inches since tilting the television at an angle would give a little more room. The Pythagorean Theorem tells you there is $\sqrt{29^2 + 80^2} = 85.1$ inches of room.

Television Size

Televisions are advertised as 32-inch, 55-inch, etc. That number refers to the diagonal length of the screen.

Examples:
1. An advertisement for a 39-inch TV states that the product width is 34⅞ inches and the product height is 20⅝ inches. Does that in fact yield a 39-

inch diagonal screen size? Using the Pythagorean Theorem actually gives a hypotenuse of 40.52 inches. Since we did not get the expected 39-inch measurement, the assumption would be that advertisement's use of "product width" and "product length" would include the border surrounding the screen.

2. A 32-inch television is produced to fit programs with a 4:3 aspect ratio. What should be the length and width of the screen?

A 4:3 ratio could be thought of as sides of 4x and 3x. Using the Pythagorean Theorem,

$$(4x)^2 + (3x)^2 = 32^2 \rightarrow 25x^2 = 1024 \rightarrow x = 6.4$$

So, the sides must have a length of (4 · 6.4 =) 25.6 inches and a width of (3 · 6.4 =) 19.2 inches.

Ladder Height

You consider purchasing a step ladder. Advertisements will state that a store has sizes of 8-, 10-, or 16-foot ladders. However, that amount is before the ladder is spread out. How tall is a 16-foot ladder really going to be when it is set up?

The on-line specifications page state that the "approximate spread" is 104 inches. A right triangle could be constructed in which the hypotenuse is 16 feet (192 inches), one leg is 52 inches (half of the full spread), and x is the actual height of the ladder when in use. Applying the Pythagorean Theorem gives the equation $192^2 = x^2 + 52^2$. Solving this equation shows the ladder to have, when in use, an actual height of just under 15 feet, 5 inches.

Circles

Of all two-dimensional shapes, only the circle has the same width any way it is measured. Manhole covers are circular, for one reason, because that is the only shape an enterprising mind could not turn so it would fall down into the sewer. The turning radius of a car and the broadcasting radius of a television station are described by circles. The shapes of orbits in electrons and solar systems are approximated by circles.

It is quite interesting how scientific discoveries were made many years before the advent of calculators and computers. The following examples are of discoveries made of the heavens thousands of years ago.

The Size of the Earth and the Moon

Eratosthenes was a mathematician, astronomer, poet, and the chief librarian at the famed Library at Alexandria. He once figured the circumference of the Earth to be 24,700 miles. We now know the actual circumference

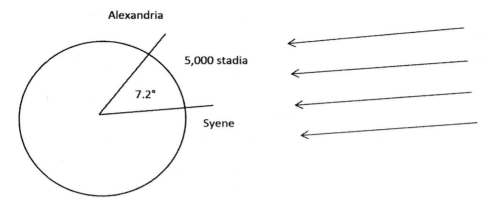

Eratosthenes measures the circumference of the Earth.

to be 24,900 miles. This is amazingly close considering it was done in 240 BC. Unfortunately, others followed his work with their own estimates, not nearly as close as Eratosthenes'. They figured the Earth as being much smaller. Using those later estimates was the main reason Columbus thought he had landed in the Orient, which was actually much farther than he believed.[15]

Not everyone in those days thought the Earth was flat. Those observing the Earth's shadow on the Moon during a lunar eclipse noted there was a rounded shadow. Eratosthenes was one of those that thought the Earth was round.

So how did Eratosthenes come up with this value? He knew the distance from Alexandria to Syene was 5,000 stadia. At the time of day when there was no shadow in Syene, there were shadows made in Alexandria of 7.2 degrees. Assuming the rays of light are parallel, Alexandria's and the Earth's central angles must be the same size. Since 7.2 degrees is one-fiftieth of a 360-degree circle, the 5,000 stadia must be one-fiftieth of the circumference of the Earth. Making a conversion gives the 24,700 mile figure.[16]

It is surprising he got that close. The precise length of one stadia varied. Also, the 5,000 stadia was an estimate based on how long it took a caravan to travel between the two cities. So, while Eratosthenes showed a great deal of creativity, he also had some amount of luck to get within 1 percent of the actual circumference. It was a very impressive feat, nonetheless.

The Size of and Distance to the Moon

Aristarchus seems to be underrated as one of the great minds of history. He was a Greek astronomer who lived in the same age as Eratosthenes. A couple of thousand years ago, he came up with distances to some of our heavenly neighbors. He wasn't always 100 percent correct, but still, pretty remarkable. For one thing, although it didn't catch on at the time, he theorized that the Earth orbited the Sun. This was fifteen hundred years before Copernicus.[17]

He found the size of the Moon. He observed the Moon as it passed through a lunar eclipse. With simple observation and some slick mathematics, he figured that an eclipse was equivalent to 3.7 Moon diameters. Thus, the Earth's diameter (thanks to Eratosthenes) divided by 3.7 should equal the diameter of the Moon.[18] It turns out that the Earth's diameter is 3.66 times the size of the Moon's.

Now, for the distance to the Moon. Viewing from the Earth, Aristarchus could measure the angle subtended by the Moon. A right triangle could be set up with vertices at the point he was standing, the middle of the Moon, and the edge of the Moon. He knew one of the acute angles and, since he had already found the diameter of the Moon, he knew one of the distances. He could then use the tangent function to find the distance to the Moon.

The Size and Distance to the Sun

First Aristarchus found the distance to the Sun. At certain times, the Moon appears to us to be half full. Aristarchus realized that, at those times, if a person was standing on the Moon, he would notice that the angle obtained by looking from the Sun to the Earth would measure 90 degrees. Now back to standing on the Earth. Aristarchus carefully measured this angle and found it to be 87 degrees. So it must be true that $\cos(87°)$ is equal to the ratio of the adjacent side to the hypotenuse, or in this case, the Moon to Earth distance divided by the Earth to Sun distance. Since he already had a Moon to Earth distance, he could solve to find the Earth to Sun distance. (He wouldn't have done these things quite as we might, because trigonometry wasn't at that stage yet, but he found other ways to find the solution.)

This time, his answer was quite a bit off. First of all, it's pretty difficult to tell when the Moon is exactly half-illuminated. If that is off, then so is the assumption of a 90-degree angle. Also, his measurement of the 87-degree angle was wrong. It turns out it is very nearly 90 degrees itself—89.85 degrees.[19]

Next, for the size of the Sun. He figured that the Sun was 19 times farther away than our distance to the Moon, and, so, 19 times the size of the Moon. He was right about the Sun being larger, but quite a bit off as to how much larger. It turns out that rather than 19 times farther away than the Moon, the Sun is more like 382 times as far away.[20]

Aristarchus wasn't always correct, but, his discoveries were quite remarkable for his time.

Constructing Circle Graphs

While most students do not have a problem interpreting circle graphs, constructing one is a bit more of a challenge. There are many sets of data that could be used in the construction of one of these graphs. This table shows the population of the Earth by continent, as of 2013, and is a good candidate for turning into a circle graph.

Continent	Population (in millions)
Asia	4,140
Africa	995
Europe	739
North America	529
South America	386
Oceania	36

A number of math skills are needed to construct this graph:

1. Find the total population.
2. Divide to find the percentage for each continent.
3. Multiply each by 360 to find the number of degrees in each sector.
4. Use a compass to draw a circle.
5. Use a protractor to draw in each sector.
6. Label each sector.

Track and Field Throwing Sectors

The shot put, discus, and hammer throw all take place in a circle. To be a fair throw in any of these events, the throw needs to be within a 34.92-degree sector from the center of the circle.[21] That seems like an odd choice for a sector size. It was chosen so it could be laid out easily by tape measure instead of dealing with trying to measure an angle.

Extend two lines from the center, both a distance of 50 feet. Position the endpoints so they are 30 feet apart. By doing this, the central angle is going to be 34.92°. This could be verified with a little trigonometry. (We'll revisit this topic later in the book.) The numbers 30 and 50 would not have to be used. Any starting distance could be used as long as the distances are in a 3:5 ratio.

The javelin event also has a sector that

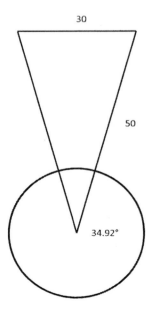

Setting up a shot put throwing area.

marks the fair throws. The measurement for its angle is 28.96°.[22] Again, ease of set-up was the determining factor in choosing this number. To set up the javelin sector, find the point 8 meters behind the foul line to be the starting point for the two lines that will serve as the boundaries of the sector. Extend these lines any distance. We'll call that amount x. Then position your lines so that the opposite side distance is one-half of x. Similar to the previous situation, this could be verified with trigonometry.

Latitude and Longitude

The geometry of circles and spheres is used in finding locations on the Earth. Any location on the Earth's surface can be found with two numbers. These two numbers are referred to as latitude and longitude. They are both based on the concept that the measurement of an arc of a circle is the same as the measurement of its subtended central angle. However, there are differences between how the values for latitude and longitude are determined.

The measurement of lines of latitude is based on a central angle whose vertex is at the center of the Earth. One side of the angle intersects the Equator. Where the other side intersects the Earth's surface determines its latitude. (We'll continue to use the word "lines," although these are actually circles.)

A great circle on a sphere is one that has the same center as the sphere's. In the setting of a sphere, this is in essence a "line" as it is the shortest distance between any two points. The Equator can be defined as that great circle on the Earth that lies in the plane perpendicular to its rotational axis. The Equator serves as a zero line when measuring latitude. A line of latitude of 10° is the circle that is an arc of 10° north or south of the Equator. The North Pole is 90°N because the central angle, and thus its arc, is 90°.

All lines of latitude are 0 to 90 degrees north or 0 to 90 degrees south of the Equator. All lines of latitude are contained in planes parallel to each other. The only line of latitude that is a great circle is the Equator.

Longitude is measured somewhat differently. While the Equator has physical determination, the zero line for longitude, the Prime Meridian, was selected somewhat arbitrarily to be the line going through the Royal Observatory at Greenwich, England. This was primarily due to the fact that a nautical almanac containing important navigational information was based on that line as a starting point. Lines of longitude then progress from 0 to 180° east or west of this Prime Meridian. Unlike lines of latitude, lines of longitude are all great circles. However, as opposed to the case for latitude, none are parallel, intersecting at both the North and South Poles. The flat maps that show parallel lines of longitude are misleading.

In addition to the Equator, there are four other lines of latitude that are important. The Earth rotates on an imaginary axis that is 23.4° from being perpendicular to its orbit. This tilt accounts for the changing seasons during the Earth's orbit of the Sun. During the summer solstice at 23.4°N and the winter solstice at 23.4° S, the Sun is directly overhead. Those lines are known as the Tropic of Capricorn and the Tropic of Cancer, respectively. Anywhere between those lines will have the Sun directly overhead at other times of the year, and locations outside of the region bounded by those lines will never have the Sun directly overhead.

There are some parts of the globe that don't see the Sun for days on end. However, those locations also have days in which the Sun doesn't set. This happens north of 66.6°N (90° − 23.4°) and south of 66.6° S. These lines are known as the Arctic and Antarctic Circles, respectively.

Rainbows

The mathematics of circles and spheres is also the basis of the rainbow. A rainbow happens when rays of light enter water droplets and reflect back to our eyes.

We often see rainbows just after a rain. At this time there are still water droplets in the air and the Sun has started to reappear from the clouds. Note that to see a rainbow, the individual has to have his or her back to the Sun. That way, the light bouncing back from the droplets can be seen. Light might be reflected away when hitting a drop of water, but some enters. The light that enters is bent (refracted) somewhat as it changes mediums from air to water. Some of this light passes all the way through the droplet, but some is reflected off the back side of the drop. When it then leaves the drop, it is again refracted as it reenters the air.[23] The various frequencies that make up white light will be refracted at different angles. The frequency associated with red light is changed the least, and violet light the most. The red light leaving these spherical droplets will emerge as a red cone shape. All the droplets will look to us to be a circle of red. The other colors will be seen as larger circles. A rainbow is actually a circle of colors that we typically only see arcs of, being cut off by the horizon.

There is no pot of gold at the end of the rainbow. One reason for this (there are others) is that there is no physical location for a rainbow. Different observers will see lights from a different set of droplets, and the rainbow will be perceived to be in a different location.

Sometimes a secondary rainbow is seen. That is because light can enter the droplet, first being refracted, then reflected at the far side of the drop, then reflected inside the drop a second time, and then finally leaving with a final refraction.[24]

With a good deal of time and work, the details of the mathematics of rainbows can be examined using a healthy amount of trigonometry and calculus.

Tangent Lines

A tangent line touches a circle, or for that matter, any curved shape, at a single point. You are twirling an object on a string around your head. The string keeps it in a circular path. If that string breaks, the object will be in flight along that tangent line from the point when the string broke. If gravity went away, the Earth would immediately head out in a straight line, tangent to its orbit. A speeding car going around a turn and hitting black ice, will leave that curve traveling in a line that is the tangent to the road at the point it hit the ice. A clothes dryer gets our clothes dry by supplying heat. But also, as the clothes spin in the drum, water droplets fly off along a tangent line and escape through the holes in the drum.

Tangent lines are an important concept in calculus. The derivative of a function can be used to find the slope of the line tangent to a curve at a given point. If a function expresses the relation between distance and time of a moving object, the slope of the curve gives the change in distance over the change in time, or the speed. The second derivative gives change in speed over the change in time, or the acceleration. The derivative of a function relating price to time gives the measure of inflation.

Earthquake Epicenters

When throwing a rock into a pond, there are waves that spread out from where the rock caused the initial disturbance. Much the same happens when an earthquake takes place. Expanding waves travel from the epicenter of the earthquake. There are primary waves (P-waves) and slower moving secondary waves (S-waves).[25]

There are many seismic recording stations throughout the world that pick up any seismic activity. It turns out only three of those locations are needed to find the location of the epicenter. Because P-waves travel faster than S-waves, the farther away from the epicenter, the farther behind the S-wave is from the P-wave. A station can record the difference between those two waves and, using that data, the distance the waves had traveled can be found.[26] A circle can then be drawn that represents all the possible locations for the epicenter. Repeating this procedure from two other stations gives two more circles. The intersection of those three circles determines the location of the epicenter of the earthquake.

Polygons

There are many examples of polygons in our world. Because of their rigidity, triangles are used to brace constructions. Pyramids are made up of triangles. Most buildings are made up of squares and rectangles. The bolts on fire hydrants are pentagons. (Regular hexagons would have opposite sides parallel, making it a little too easy for non-firemen to turn on the water.) The Pentagon is a pentagon. Pencils are often hexagonal so they won't slide off desks. Honeycombs are made of hexagons because hexagons tessellate. Stop signs are octagons. Crystals are composed of various polygons. Scientists classify crystals by grouping them into seven different polygonal shapes— cubic, orthorhombic, hexagonal, etc.[27] Molecules are described as being polyhedral in nature. There are types of viruses that can be classified as a polyhedral virus.[28] Even the pretend world is polygonal. Surfaces in video games and computer generated movies may appear smooth, but are made up of polygons. In the early stages, it was often easy to make out the edges of the polygons. With enhanced technology, those polygons have become so small they are virtually undetectable.[29]

The Value of Pi

The human race has known for a long time that the distance around a circle is about three times the distance across it. Over two thousand years ago, Archimedes decided to attempt to find that ratio by using regular polygons. He could find the perimeter of a regular polygon without too much difficulty. Inscribing and circumscribing regular polygons inside and outside a circle made for good approximations for the circle itself. The more sides his polygons had, the closer they were to being circles. To use one of these polygon's perimeter and divide it by the distance across should give a value close to pi. He ultimately used a 96-sided polygon. By doing so, he pinned down pi to being between 3.1408 and 3.1429.[30]

Geodesic Domes

Geodesic domes are structures that are dome shaped and made up of many polygons; often triangles. A rectangle of a particular length and width can be morphed into other parallelograms with the same lengths of sides. Not so with triangles. By the SSS postulate, any triangle has a fixed shape, making it, by architect R. Buckminster Fuller's thinking, an excellent design shape for his geodesic domes. Calculus shows the spherical shape of domes to have less surface area per volume than other shapes, making it much more efficient in minimizing heating and cooling costs and using fewer building materials.

The large building at Epcot Center is an example of a geodesic sphere. It is made up of 3,840 triangles. These triangles are turned into pyramids, with three triangles attached to those bases. This makes the Epcot dome an 11,520-gon polyhedron.[31]

Icosahedrons

An icosahedron is any polyhedron with twenty faces. A soccer ball can be said to be a truncated icosahedron.

Buckminster Fuller used icosahedrons as a basis for his geodesic dome design. Truncated icosahedrons are also found on the surfaces of a large number of viruses. The shape is also that of the carbon molecule named after him, Fullerene (C_{60}).[32]

Center of Mass

The geographic center of the 48 contiguous states is at a point just south of the Kansas-Nebraska border at 39°50' N and 98°35' W. Including Alaska and Hawaii moves the geographic center some 500 miles to the northwest. The center of population, because of the large number of people on the eastern seaboard, is in the middle of Missouri.

The center of mass of a body is an important concept in physics. Many problems in physics can be more easily handled by considering the point that is the center of mass rather than the entire body. As a high jumper goes over the bar, even though arms and legs are shifting positions, the center of mass follows a parabolic shape. Because of the arched back high jumpers employ, for good high jumpers, their center of mass actually travels under the high jump bar.

The Earth is said to orbit the Sun. In a way, the Sun is also orbiting the Earth. Both objects are affected. They are both orbiting some point which is their collective center of mass. That center of mass can be found. If two bodies are not moving in a path they would be predicted to take, there is likely a third body whose gravitational force is affecting that center of gravity and those orbits. This concept is what has allowed scientists to predict that a previously undiscovered planet must be in a certain location before it was ever seen.

This concept is seen in the centroid of a triangle. The three medians of a triangle intersect in one point called the centroid. This is the center of mass of the triangle. If a person cuts out a triangle, the centroid is the point in which the triangle balances.

There are three other points of concurrency in any triangle. All are useful in some way.

The intersection of the three altitudes is called the orthocenter. An altitude is a line segment with one endpoint at a vertex of a triangle and meets the other side at 90°. For the triangle area formula, $A = \frac{bh}{2}$, the length of any of those altitudes could be used as the value of h in the formula. That height, along with its corresponding base, could be used to find the area of the triangle.

The intersection of the three perpendicular bisectors is called the circumcenter. That point is equidistant from each triangle vertex. So, in order to inscribe the triangle in a circle, it is necessary to first construct the circumcenter.

The intersection of the three angle bisectors is called the incenter. The incenter would serve as the center of the circle that could be inscribed inside the triangle.

Perimeter, Area and Volume

It is not too difficult to find applications involving perimeter, area, and volume. This section will present several, but there are many other good applications as well.

Odometers

Odometers display how far a vehicle has traveled based on the number of times the tires rotate. This was previously done mechanically with a cable and a series of gears. Now there are digital odometers that act as miniature computers, but the basic idea remains the same.

A trundle-wheel works on the same concept. It is simply a wheel with a handle that a person rolls along the ground. It records distance based on the size of the wheel. It can be used to measure distances over various terrains, and can be used for activities such as measuring the length of a cross-country course.

Example:

A bicycle's tire has a diameter of 26 inches. How many times must it revolve to have an odometer turn over one-tenth of a mile?

There are 5,280 feet in a mile and 12 inches in a foot, making 63,360 inches in a mile. That means there are 6,336 inches in one-tenth of a mile. In one revolution, the bike tire covers $26\pi = 81.68$ inches. So, $6,336 \div 81.68 = 77.57$ turns of the wheel.

Related questions could be asked.

How many revolutions are necessary to accumulate 3.1 kilometers?

What size wheel is necessary so that 1,200 revolutions would equal one mile?

Treadmills

A treadmill is calibrated so the amount that the running surface, or belt, moves is registered on a display for the exerciser to view. How many revolutions of the belt would it take to travel one mile?

First the length of the belt would have to be found. One way is to find the length of the loop the belt makes by using geometry. The belt travels on two rollers. Thus, it is two half circles and two straight distances. To find the distance around would be to find πD + 2L, where D is the diameter of the roller the belt runs on and L is the length of each straight distance. If there is a belt that has a value of D being 2 inches and each straight distance is 52, then the length of the belt is 110.28 inches. How many belt revolutions would it take to register one mile?

$$5{,}280 \cdot 12 \div 110.28 = 574.5 \text{ revolutions}$$

Airplane Lift

Several things work together to determine the amount of lift an airplane receives. An important part of that determination is the surface area of the wing. The following formula can be used to determine the amount of lift[33]:

$$\text{Lift} = c \cdot \frac{dv^2}{2} \cdot A$$

C is a coefficient usually measured experimentally and is influenced by the shape and angle of attack of the wing. The variable d is the density of the air (slug/ft³), v is the plane's velocity relative to the air (ft/sec), and A is the area of the wing (ft²).

Examples: The following two problems are supplied by NASA.[34]

1. Calculate the lift generated by an aircraft flying at 10,000 feet at a speed of 120 ft/sec. The wing area is 500 square feet, the lift coefficient is 1.67, and the density is 0.00165 slug/ft³.

[Answer: 9,919.8 pounds]

2. Find the velocity of an aircraft that generates 30,590 pounds of lift when the wing area is 800 square feet. The lift coefficient for the aircraft is 1.4, and the air density is 0.00089 slugs/ft³.

[Answer: 247.7 ft/sec]

Paint Coverage

The label on a can of paint will state how many square feet the paint would be expected to cover. This is helpful information. If the painter buys too many cans, that will be costly. Paint stores will virtually never take back paint that has already been mixed. Less costly, but more of a hassle, is having to go back to the paint store every time another gallon of paint is needed. The best method is to figure out the square footage ahead of time.

Example:

The label on a gallon of paint states it will cover 400 square feet. How many gallons might be needed for the walls of a room that are 8 feet tall and has a floor and ceiling that are 20 by 30 feet?

[Answer: 2 gallons]

There are other situations that are similar to this painting problem. Lawn fertilizer is to be applied for a certain amount per so many square feet. To be well over that ratio will harm, if not kill, the grass. To be under that ratio will likely be ineffective.

Circle Irrigation

Perfectly formed circles can be seen when flying over much of the United States. As opposed to the alien created crop circles, these are formed by circular irrigation. Where irrigation takes place, crops grow. Circle irrigation, or center pivot irrigation, involves a pipe extending from the pivot point. That pivot point is the center and the length of pipe is the radius of the circle. Water flows out of several nozzles located along the length of the pipe. Those nozzles increase in size the farther out they are from the center. This allows for an equal distribution of water. Although not all of the land would be irrigated, this method has the advantage of not needing as many people to operate as other methods.

Examples:
1. If the pipe is 380 yards in length, how many square yards could be watered?
$$A = \pi \cdot 380^2 = 453,646 \text{ square yards}$$
2. To water 80 acres, what would the length of pipe need to be? (There are 4,840 square yards in an acre.)
$$80 \cdot 4,840 = \pi r^2 \rightarrow r = 351 \text{ yards}$$
3. A rectangular piece of land that is 400 yards by 1,200 yards is going to be irrigated by center pivot irrigation. Three pipes each with a length of 200 yards are going to be used. What percent of the land will be irrigated?

This would be found by finding the area of the three circles and dividing by the area of the rectangular piece of land.
$$3(\pi \cdot 200^2) \div (400 \cdot 1,200) = 0.785 = 78.5\%$$

Pizza Costs

Here are some actual prices for pizza taken from a local eatery.

An extra-large 16-inch (diameter) pizza—$21.75
Large 14-inch pizza—$19.75
Medium 12-inch pizza—$16.45
Small 10-inch pizza—$12.45
Mini 8-inch pizza—$8.20

If they compute the cost per square inch, students might be surprised that not all pizzas are the same. The extra-large costs 10.8 cents for every square inch of pizza. On that basis, it's the bargain. They might also be surprised that the 16-inch is not twice as much pizza as the 8-inch, but actually four times as much.

Speed of the Earth

The Earth's orbit around the Sun is not a perfect circle, although it is fairly close. Students could use the formula for the circumference of a circle to answer a couple of questions.

How fast is the Earth orbiting the Sun in miles per second? Given the fact that the Earth is 93,000,000 miles from the Sun and takes a year to complete its orbit, this could be found without too much work.

$$\frac{2\pi(93,000,000)}{365.25(24)(60)(60)} = 18.5 \text{ miles per second}$$

The Earth itself is very close to being a perfect sphere. How fast is a person moving that is standing at the Equator as the Earth rotates? The radius of the Earth is 3,959 miles and a rotation is completed each day.

$$\frac{2\pi(3,959)}{24(60)(60)} = 0.29 \text{ miles per second}$$

Building a Track

Construction is to begin on a 400-meter track. It is to have two curves of 100 meters each and two straightaways. What should be the radius of the curves?

The two curves together make one circle with a circumference of 200 meters. Since $2\pi \cdot r = 200$, the radius of the circle is 31.831 meters.

That issue is solved. However, the distance around the track is 400 meters only if a runner is in the first lane. Running in the outer lanes will mean running a greater radius, thus a greater circumference. The solution to this is to find that added distance for each lane and allow those lanes to have what appears to be a head start, but really just evens things up. This distance is referred to as the stagger.

According to the NCAA, "The width of each lane shall be constant and not less than 1.067 meters. It shall normally consist of two parallel straights and two semicircular curves of equal design."[35]

Let us make our track with lanes that are 1.2 meters in width. The distance from the center of the half circle to the inside of lane one has already been found at 31.831 meters. The distance for the center of the half circle to the inside of lane two is (31.831 + 1.2 =) 33.031 meters. This means that the lap distance, if running in lane two, is 200 + 2π(33.031) = 407.54 meters. Thus, the stagger needs to be placed 7.54 meters ahead in lane two, and then in multiples of that amount for each succeeding lane.

Integral Calculus

Integral calculus can be used to find the area under a curve. The area can be found with rectangles fitted underneath the curve. The exact integral is found by a process in which the width of those stacked rectangles get smaller and smaller (i.e., the limit approaches zero.) The trapezoid method is similar, but is an improvement on stacking rectangles as the trapezoid gives a better fit to the changes in height of the curve.

Famous Structures

There are a number of well-known structures that can be used as sources of applications. Finding the area of one side of a structure, the distance to walk around it, the surface area, the volume, the cost of painting the object, the length of a diagonal on one face, using trigonometry to find angles of elevation—all are possible applications. A few sample questions will be provided, but many more potential questions are available. (Sometimes different sources will state slightly differing measurements than those given here. This is especially true in some of the ancient buildings as decay has made estimates necessary.) There are many other examples of buildings that can be used, but here is a start.

The Roman Colosseum

The Colosseum was constructed between AD 70 and 80 and was the scene of great entertainment including gladiator battles and animal fights. It could seat over 50,000 spectators. It has an oval shape, 620 by 512 feet. There is some debate whether the Colosseum is simply an oval or whether the Romans possessed the mathematics necessary to design it as an ellipse. In any event, it is close. The interior floor, where the action took place, is also an oval, and is 287 by 180 feet. The height of the Colosseum is 157 feet.

1. Are the interior arena oval and the Colosseum building, similar shapes?

2. Make scale drawings of the building including the inside arena.

3. Assuming the floor and the entire stadium are ellipses, find the areas. (The area of an ellipse is pi multiplied by half the major axis by half the minor axis.)

4. Using the previous answer and assuming a sell-out crowd, how many square feet would be available per spectator?

5. If a gallon of paint covers 400 square feet, how many gallons would it take to paint the floor of the Coliseum?

The Great Pyramid

It was built in 2550 BC probably as a tomb for pharaohs, although this is uncertain. Each side of its base is 755.8 feet. Its original height is estimated to have been 480.7 feet. Students can debate whether it was by design or coincidence that the perimeter divided by the height is extremely close to the value 2π.

1. What is the slant height of this pyramid?
2. What is the length of a lateral edge?
3. Find the area of a lateral face.
4. The building blocks of the pyramids varied in size, but averaged 5 by 5 feet. If that is the case, estimate the number of blocks that are on a lateral face.
5. What is the volume of the pyramid?

The Parthenon

The Parthenon in Athens, Greece, was a temple dedicated to the goddess Athena. It was completed in 438 BC. The top step of the temple is 101.4 by 228 feet. The columns of the Parthenon are 6 feet 2 inches in diameter and are 34 feet 3 inches tall.

1. What is the lateral area of one of the columns?
2. What is the area of the top step of the Parthenon?
3. An expensive clear coating is to be used to protect the top step. If its cost is $1.50 per square foot, what is the total cost?

The Dome of the Rock

The Dome of the Rock was erected in Jerusalem in AD 691. The dome itself is approximately a half of a sphere with a diameter of 66.3 feet. It sits upon a solid base that is a regular octagon. The walls supporting the octagon are rectangles that are 60 feet long and 35.5 feet tall.

1. What is the volume of the hemisphere?
2. Using trigonometry, find the length of the of the octagon's apothem.
3. What is the area of the octagonal base?

The Leaning Tower of Pisa

In 1173, work was done on what was probably first known simply as the Tower of Pisa. It served as a bell tower for the nearby cathedral. It was designed to stand 185 feet tall. The tower has a diameter of 50.8 feet. A spiral walkway to the top surrounds a hollow cylinder that has a diameter of 14.75 feet. Even while being built, the tilt was beginning. Although it used to be more pronounced, work has been done so the lean is now 4 degrees.

1. What is the outside surface area of the tower?
2. Excluding the hollow interior, what is the volume of the tower?
3. Make a scale drawing of the tower with its lean of 4 degrees.

The Arc de Triomphe

Work on the arc, a memorial to French war dead, was completed in 1836. It has a height of 162 feet, a width of 150 feet, and a depth of 72 feet. There are four large vault cutouts in the sides of this monument. Each one is a rectangle topped with a semicircle. Each rectangle's height is 95.8 feet and its width, and thus also the diameter of the semicircle, is 48.8 feet. The smaller pair of vaults have heights of 61.3 feet and widths of 27.7 feet.

1. Find the area of the largest cutout opening.
2. Find the surface area of each face.
3. What is the area of the top of the Arc?

The Space Needle

The Seattle Space Needle, built for the 1962 World's Fair, reaches to 605 feet. A restaurant is 500 feet above the ground. It has an outer diameter of 94.5 feet. The restaurant rotates and originally took 58 minutes to make one lap. It has been sped up and now takes only 47 minutes. It is a 41 second elevator ride for 520 feet up to the observation deck. It rests on a concrete foundation that is 120 by 120 by 30 feet.

1. How many cubic feet of concrete was used in the foundation?
2. How fast is the outer edge of the restaurant moving in feet per minute?
3. How fast is the elevator going in miles per hour?

Vietnam Memorial

The memorial was designed by Maya Ying Lin, an undergraduate at Yale University, and was built in 1982. At its greatest height, it is 10 feet 3 inches. The top is at ground level, so the memorial itself actually sits below ground. Each wall is 246 feet, 9 inches long. The walls meet at an angle of 125°12⊠. Its design is such that one wing points directly to the Washington Monument and the other to the Lincoln Memorial. What seem to be giant triangles are actually trapezoids with a long base of 10 feet 3 inches, tapering to small bases of 8 inches.

1. Using the Law of Cosines, how far is it from one end to the other?
2. What is the area of the memorial walls?
3. What is the slope of the walkway at the base of the memorial?

The Louvre Pyramid

A glass and metal pyramid sits in front of the Louvre Museum. Designed by I.M. Pei and constructed in 1989, it consists of almost 700 glass rhombuses and triangles. It has a height of 70.9 feet and has a square base with sides of 115 feet.

1. What is the pyramid's slant height?
2. What is the lateral area?
3. What is the volume?

The London Eye

This giant observation wheel in the middle of London opened in 2000. It reaches a height of 434 feet above ground level. The wheel itself has a diameter of 424 feet. It takes half an hour to complete one revolution.

1. What is the circumference of the wheel?
2. How fast is the wheel turning in feet per minute?
3. How does the speed of the London Eye compare with the speed of the restaurant at the top of the Seattle Space Needle?

III

ALGEBRA II

Solving Quadratic Equations

Many of the following problems require the use of the quadratic formula or completing the square. Despite what is found in most algebra textbooks, most quadratic expressions don't conveniently factor. However, some numbers could be changed so that they do, making them useful in chapters dealing with factoring polynomials.

Work Problems

Zack and Lucy are painting a barn. It takes them 5 hours working together. If they were working alone, they estimate it would have taken Lucy about an hour longer to paint the barn than Zack. If so, how long would it take Zack or Lucy if they attempt to paint the barn alone?

Let x be the number of hours it would take Zack to paint the house alone. Then the rate Zack can paint a house per hour is 1/x. Lucy is working at a rate of $1/(x + 1)$ houses per hour. The rate multiplied by time would give the amount of work each would do on the job. So completing the total job would call for solving:

$$5\left(\frac{1}{x}\right) + 5\left(\frac{1}{x + 1}\right) = 1$$

To solve this, each side could be multiplied by $x(x + 1)$ yielding a second degree equation, $x^2 - 9x - 5 = 0$, which can be solved with the quadratic formula. Doing so gives solutions of x = –0.52 and 9.52. To answer the original question, the positive value is taken, and so it would take Zack 9.52 hours and Lucy 10.52 hours.

The Golden Ratio

Mankind has been aware of the golden ratio for over 2,000 years. It has made its appearance in nature, architecture, and music. It can be defined as a proportion. Divide a line segment of a length x + 1 into lengths of x and 1. The ratio of the whole line segment of length x + 1 to the segment of length x is the same as the ratio of line segment of length x to the segment of length 1. This is a proportion which can be written as $\frac{x+1}{x} = \frac{x}{1}$. The resulting quadratic, $0 = x^2 - x - 1$, gives solutions of $\frac{1 \pm \sqrt{5}}{2}$. Of these two solutions, the positive value getting is the golden ratio: $\frac{1 \pm \sqrt{5}}{2} \approx 1.618$.

Boat Trip

Gilligan takes passengers on a three-hour tour downriver for 70 miles and then back to the original starting point. He goes downstream with an 8-mile-an-hour current, but then has to come back against that same current. What would be his speed in still water?

Since $d = r \cdot t, t = \frac{d}{r}$, going downstream would take $\frac{70}{r+8}$ hours, and upstream would take $\frac{70}{r-8}$ hours. An equation showing the round trip is $\frac{70}{r+8} + \frac{70}{r-8} = 3$.

This eventually yields the equation $0 = 3r^2 - 140r - 192$, which can again be solved with the quadratic formula or by factoring. The answers are r = $-\frac{4}{3}$ and 48. Therefore, the boat would travel at 48 mph in still water.

Running

A runner goes north at a 4-mile-per-hour pace, and another goes west an hour later at 5 miles per hour. When will they be 10 miles apart? Using the formula d = rt, after t hours, they will have gone 4t and 5(t – 1) hours, respectively. By the Pythagorean Theorem, $10^2 = (4t)^2 + (5(t-1))^2$.

$$10^2 = (4t)^2 + (5(t-1))^2 \rightarrow 0 = 41t^2 - 50t - 75 \rightarrow t = 2.093, -0.874$$

In 2.093 hours, or 2 hours 6 minutes, they will be 10 miles apart.

Falling Bodies

Galileo showed that objects, regardless of mass, fall at the same rate. While there is some debate whether this took place at the tower in Pisa or elsewhere, there is no debate about how important this discovery was. Without air getting in the way, a sheet of paper would fall at the same rate as a book. In fact, an easy demonstration of this is to place the paper on top of the book and then drop them. The following formula describes this motion and can generate many examples of quadratic equations.

$$y = \frac{-gt^2}{2} + v_0 t + y_0, \quad \text{where } t = \text{time (in seconds)},$$

v, v_0 = current velocity and initial velocity (meters per sec),
y, y_0 = height after t seconds and initial height (meters).

The acceleration due to gravity is g, which on Earth is 9.8m/sec². This formula can also be used with English system units by using 32 ft/sec² as the value for g. This formula can also be used to determine the motion of falling objects on other planets. On tiny Mercury, g = 3.61 m/sec². On enormous Jupiter, g = 26.0.

Examples:

1. Your hand is 2 meters off the ground when it throws a ball 5 m/s into the air. How long will it take to land?

This can be found with the equation: $0 = -4.9t^2 + 5t + 2$.

Answers: t = −0.31, 1.33. As is often the case with these types of problems, only the positive value is taken, so the answer is 1.33 seconds. If looking at the complete parabola, both values are solutions for t, but the ball was released at t = 0, thus the −0.31 isn't part of the domain.

2. How long does it take a ball dropped from 120 meters to hit the ground?

$0 = -4.9t^2 + 0t + 120$ [Answers: t = ±4.95]

3. How long would it take to hit the ground if the ball is thrown straight up at 10 m/s from 120 meters?

$0 = -4.9t^2 + 10t + 120$ [Answers: t = 6.07, −4.03]

4. How long would it take if the ball is thrown *downward* at 2 m/s?

$0 = -4.9t^2 - 2t + 120$ [Answers: t = 4.75, −5.16]

5. A rock is dropped from a spaceship 52 meters above the surface of Jupiter. In how many seconds will it hit the ground? While we're at it, how fast was it going?

$0 = -13t^2 + 0t + 52 \rightarrow 0 = -13(t - 2)(t + 2)$ [Answer: t = 2, −2]

The rock will hit the ground in 2 seconds. To find its speed, the formula $v^2 = v_0^2 + 2gt$ can be used. So, $v^2 = 0 + 2 \cdot 26 \cdot 2 = 104$. The rock's speed when it lands on Jupiter is $\sqrt{104} = 10.2$ meters per second.

There are many formulas that can give additional information about the path of projectiles:

$$v = v_0 + gt, \qquad v^2 = v_0^2 + 2gt, \qquad d = \tfrac{1}{2}(v_0 + v)t, \qquad d = v_0 t + \tfrac{1}{2}gt^2$$

Braking Time

A more generalized version of the above equation can be used for motion other than that influenced by gravity: $y = \dfrac{at^2}{2} + v_0 t + y_0$. This time the value of "a" is whatever the rate of acceleration may be for the object.

Examples:

1. A car is traveling at 45 miles per hour (66 ft/s) and brakes to a stop in 80 feet, 5 seconds later. What was the car's acceleration?

$$80 = \frac{a \cdot 5^2}{2} + 66 \cdot 5 + 0$$

Solving this equation finds the acceleration of the car to be –20 ft/s². The value is negative because the car was decelerating.

2. On a dry road, a car can safely decelerate at 15 ft/s². If that is the case, how many seconds would it take a car going 60 mph (88 ft/s) to travel 100 feet?

The answer can be found with the equation $100 = \frac{-15t^2}{2} + 88t + 0$. Using the quadratic formula gives two answers, 1.27 and 10.46 seconds. The car would stop in 1.27 seconds.

Braking Distance

To a great degree, a car's stopping distance is determined by the friction between its tires and the road. The formula used to determine the minimal stopping distance is[1]: $D = \frac{v_0^2}{2\mu g}$, where D is the distance (meters), v_0 is the velocity at the start of the braking (meters per second), g is 9.8 m/sec², and μ is the coefficient of friction. The value of μ is approximately 0.7 for dry roads, 0.4 for wet roads, and 0.2 for icy roads.

Additionally, there is some reaction time that passes before the brakes are applied. This time can vary, but is typically taken to be 1.5 seconds. This added amount makes our total braking distance: $D = 1.5v_0 + \frac{v_0^2}{19.6\mu}$.

For any given stopping distance and coefficient of friction, the velocity of the car could be estimated by solving this quadratic equation.

Examples:

1. If a car traveling on a dry road takes 65 meters to stop, approximately how fast was its speed before the brakes were applied?

Substituting values gives the equation: $65 = 1.5x + \frac{x^2}{19.6(0.7)}$ $\rightarrow$ $0 = \frac{1}{13.72}x^2 + 1.5x - 65$

The initial velocity turns out to be 21.296 meters per sec or 47.6 miles per hour.

2. If a car is on icy roads and takes 100 meters to come to a stop, what is an estimate for how fast the car was traveling? [Answer: 17.08 m/sec (38.2 miles per hour)]

Exponents

Scientific Notation

Scientific notation is especially helpful in writing very large or very small numbers. A knowledge of exponents is necessary to make sense of them and then to compute with them.

Neurons in the human brain: 1×10^{11}
Connections between neurons in the brain: 1×10^{14}
Newton's Universal Gravitational Constant: 6.67×10^{-11}
Avogadro's Number (number of atoms in 1 mole of an element): 6.02×10^{23}
Diameter of an atom of carbon: 2.2×10^{-8} cm
Volume of the Sun: 1.41×10^{33} cm^3

Here are some examples of calculations involving the Sun:

1. How many carbon atoms could fit inside the Sun?

$$\frac{1.41 \times 10^{33}}{2.2 \times 10^{-8}} \approx 0.64 \times 10^{41} = 6.4 \times 10^{40}$$

2. What is the approximate diameter of the Sun?
Volume of a sphere = $\frac{4}{3} \pi r^3$ → $1.41 \times 10^{33} = \frac{4}{3} \pi r^3$ → $r = 0.696 \times 10^{11}$
Doubling this value gives a diameter of 1.392×10^{11} centimeters.

3. What is the diameter of the Sun in kilometers?
There are 100 centimeters in a meter and 1000 meters in a kilometer. So,
$\frac{1.392 \times 10^{11}}{10^2 \cdot 10^3} = 1.392 \times 10^6$ kilometers.

4. How far is a light-week?
Most are familiar with the term light-year. How far can light travel in a week? Light travels at 186,000 miles per second. There are 7 days · 24 hrs/day · 60 min/hr · 60 sec/min = 604,800 seconds in a week. Since distance is equal to rate times time, there are approximately $(1.86 \times 10^5)(6.048 \times 10^5) = 11.25 \times 10^{10}$ = 1.125×10^{11} miles in a light-week.

5. How long does it take light to go from the Sun to the Earth?
The distance is 93 million miles and light travels at 186,000 miles per second.

$D = r \cdot t \rightarrow 9.3 \times 10^7 = (1.86 \times 10^5)t \rightarrow t = 5.0 \times 10^2$

So, it takes 500 seconds or 8 minutes 20 seconds

Planck's Constant

The amount of energy contained in a photon of light can be found with the equation $E = \frac{hc}{\lambda}$, where E is the energy in joules, c is the speed of light in m/s (3.00x10⁸), h is Planck's constant (6.6262x10⁻³⁴), and λ is the length of a wavelength of that light in meters.

Examples:
1. The amount of energy in one photon of yellow light is
$$E = \frac{(6.6262 \times 10^{-34})(3.00 \times 10^8)}{5.8 \times 10^{-7}} = 3.37 \times 10^{-19}.$$
2. If a photon has 1.2x10⁻²⁰ joules of energy, what is its wavelength? [Answer: 16.57x10⁻⁶ meters]

Hard Drive Space

Strictly speaking, a kilobyte would be 1,000 bytes. However, in some applications, they are considered to be 1,024 bytes. This number comes from the fact that computers operate on a base two system in which numbers are recorded as a series of ones and zeros. The fact that the value of 2^{10} is equal to 1,024, made this a convenient measurement. However, for the sake of simplicity, a kilobyte is often simply considered to be 1,000 bytes.[2] The following shows various relationships regarding computer storage amounts:

1 kilobyte (KB) = 1,000 bytes
1 megabyte (MB) = 1,000 kilobytes
1 gigabyte (GB) = 1,000 megabytes
1 terabyte (TB) = 1,000 gigabytes

Examples:
1. A computer with 1.2 GB of storage space has how many more times the space than one with 500 MB of space?

$$\frac{1.2 \times 10^9}{500 \times 10^6} = 0.0024 \times 10^3 = 2.4 \text{ times the space}$$

2. A terabyte is how many bytes?

$$10^3 \cdot 10^3 \cdot 10^3 \cdot 10^3 = 10^{12} \text{ bytes}$$

Similarly, metric conversions using exponent properties can be handled by knowing how to compute with exponents.
3. Two centimeters is what fraction of 5 kilometers?

$$\frac{2 \times 10^{-2} \text{ meters}}{5 \times 10^3 \text{ meters}} = 0.4 \times 10^{-5} = 0.000004$$

Musical Frequencies

Play the A above middle C on the piano. Looking at the graph of its sound wave on an oscilloscope shows a frequency of 440 cycles per second or 440 hertz. Twelve keys up on the piano is another A, this time an octave higher. It has a frequency of 880 hertz. The next A is 1,760 hertz. This doubling or halving effect is true for any notes an octave apart.

Pythagoras developed the musical scale made up of ratios. He found that the lengths of strings emitting the tones of a scale followed a pattern. (This was later found to be true of their associated frequencies as well.) A string playing the second note of the scale was a ratio of 9:8 to that of the first note. The third note was in a ratio of 81:64. This continues with ratios for each note of the scale until the eighth note of the scale, a full octave higher, was in the ratio of 2:1.[3]

Some 2,000 years later, around the time of Johann Bach, it was decided to make the ratio between each note the same. This is called an equal tempered scale. They wanted to keep the ratio between each successive note the same and still have the octaves be in ratio 2:1. Including half-notes, there are twelve notes from one note to the same note an octave above. Since $\left(\sqrt[12]{2}\right)^{12} = 2$, the key ratio becomes $\sqrt[12]{2}$. What about the black key above the original A? Its frequency is $440\sqrt[12]{2}$. To find the frequency of any key on the piano, use the formula $f = 440 \cdot 2^{\frac{n}{12}}$, where n is the number of notes above or below A.[4]

These two different approaches give fairly similar frequencies. The note A has a frequency of 440 hertz. The next note of the A major scale is B. Using the Pythagorean ratio of 9:8 would have B with a frequency of 495. The note B is also 2 half-steps above A. Using the even tempered scale would give 440 $\cdot (\sqrt[12]{2})^2 = 493.88$ hertz. They do have different frequencies, but probably only musically trained ears would really sense a difference in pitch.

This particular application gives a good physical example of zero, fractions, and negative exponents. Again, using the note A with a frequency of 440 hertz:

A is $440 \cdot 2^{\frac{0}{12}} = 440$ hertz

C above A (up three piano keys or half-notes) is $440 \cdot 2^{\frac{3}{12}} = 440\sqrt[4]{2} = 523.25$ hertz

A flat (down one key) is $440 \cdot 2^{\frac{-1}{12}} = \frac{440}{\sqrt[12]{2}} = 415.30$

Dimensional Analysis

The rules of exponents can be used to make sure that the units in an applied problem are correct. For example, the formula for the escape velocity from a planet is $V = \sqrt{\frac{2Gm}{r}}$, where G is the gravitational constant which is labeled $m^3 \cdot kg^{-1} \cdot s^{-2}$, m is the mass in kilograms, and r is the radius in meters. Does this really end up with a measurement of velocity? Looking only at the units themselves, the formula would give: $V = \sqrt{\frac{2Gm}{r}} = \sqrt{\frac{(m^3 \cdot (kg)^{-1} \cdot s^{-2})kg}{m}} = \sqrt{\frac{m^2}{s^2}} =$ meters per second, the correct label for velocity.

Solving Equations Containing Exponents

3D and 4D Pythagorean Theorem

The Pythagorean Theorem can be written as $d^2 = x^2 + y.^2$ This is a good application of an exponential equation. It can get more involved, however. A z^2 term can be added on the equation for finding distances through a three dimensional space. There is a Pythagorean Theorem used in space-time physics, although things get a bit more complicated here. The formula $d^2 = x^2 + y^2 + z^2 - (ct)^2$ finds a space-time interval (c is the speed of light).[5] The formula relates the three spatial dimensions with time. It is in this sense that time is considered the fourth dimension.

Pythagorean Theorem of Baseball

A baseball team is halfway through its season. They have won 60 percent of their games. That team is clearly pretty good. However, there is no guarantee that they will continue that pace and win 60 percent of their games in the second half of the season. Bill James, famed baseball statistician, developed what is called the Pythagorean Theorem of Baseball to give a better estimate of how well a team is truly doing. It is based upon the number of runs the team has scored and the number it has given up during the season. The winning percentage (P) is approximated to be[6]:

$P = \frac{S^2}{S^2 + A^2}$ where S = runs scored by offense and A = runs given up by defense.

Examples:

1. In 2014 the Oakland A's were the best team in baseball for the first half of the season. They had 51 wins and 30 losses. The second half was a different story. They won 37 and lost 44. So how good were the A's? Substituting their season's runs scored for and against into the Pythagorean Theorem of Baseball their winning percentage would be:

$P = \frac{729^2}{729^2 + 572^2} = 0.619$

Using that number for the 162 game season yields a record of 100 and 62. Their overall actual record of 88 and 74 would imply they underperformed, at least the second half of the season. In fact, that percentage would imply a half season of 50 and 31, which compares well to their actual first half of 51 and 30. Thus, the first half of their season is probably more indicative of the team they really were than their second half.

2. Suppose you know what kind of pitching your team will have this season. You figure you'll give up 70 runs. How many runs does the team need to score to win 80 percent of their games?

$$0.8 = \frac{x^2}{x^2 + 70^2}$$

The answer turns out to be 140 runs scored. Interesting. Does this imply that a team needs to score twice as many runs as it gives up if it plans on winning 80 percent of its games? To know, students could solve the equation $0.8 = \frac{x^2}{x^2 + y^2}$ for y. [Answer: Yes.]

Bill James has tweaked his original formula by using 1.83 as the exponent. While more interesting mathematically, it gives pretty much the same results. Using the new exponent gives the above mentioned Oakland A's season a predicted win total of 99 rather than 100.

The new and improved formula[7] is $P = \frac{S^{1.83}}{S^{1.83} + A^{1.83}}$

Stepping it up mathematically, another analyst feels the key to finding this mystery exponent should be found with the following[8]:

Exponent $= 1.51 \cdot \log_{10}\left(\frac{S+A}{162}\right) + 0.44$

Others have adapted similar formulas for basketball and football. For example, the following has been used to predict the winning percentage for football teams[9]:

$$P = \frac{F^{13.91}}{F^{13.91} + A^{13.91}}$$

The percent of wins would be found by substituting in the points scored for (F) and against (A) your team.

There are many other predictive formulas. To try out the formulas, points and runs for teams can be found in newspaper sports pages and on internet websites.

Volume/Surface Area

Many problems involving volume and surface area depend on being able to solve equations containing exponents. For example, the volume of a sphere is found with the formula $V = 4/3\pi r^3$ and its surface area is $A = 4\pi r^2$.

Examples:

1. The volume of the Earth is 2.6×10^{11} cubic miles. What is the surface area of the Earth?

$$2.6 \times 10^{11} = \frac{4}{3}\pi r^3 \quad \rightarrow \quad \frac{1.95 \times 10^{11}}{\pi} = r^3 \quad \rightarrow \quad r = 3,959 \text{ miles}$$

$A = 4\pi(3,959)^2 = 1.97 \times 10^8$ square miles

2. You read that the U.S. comprises 1.9 percent of the surface of the Earth. If this is true, what is the approximate area of the United States?

$(1.97 \times 10^8)(1.9 \times 10^{-2}) = 3.743 \times 10^6 \approx 3,743,000$ square miles

Kinetic Energy

Kinetic energy is an important concept in physics. The amount of energy in a moving object is given by $E = \frac{1}{2}mv^2$, where energy is measured in joules, mass is in kilograms, and velocity is in meters/second.

Examples:

1. An 850 kilogram compact car is traveling at 50 miles per hour. What is its kinetic energy?

The car's speed is 50 miles per hour or 22.35 meters per second, so E = $\frac{1}{2} \cdot 850(22.35)^2 = 212,297$ joules.

2. If that car was a 5,000 kilogram truck, what speed would it be traveling to have the same kinetic energy as the car?

This gives an equation of $212,297 = \frac{1}{2}(5,000v^2)$, and a much slower speed speed of 9.22 m/s or 20.62 mi/hr.

Roller Coaster Speed

The law of conservation of energy states that the total energy of a system cannot change. So, as an object travels, its kinetic energy, measured by the formula $KE = \frac{1}{2}mv^2$, and its potential energy, PE = mgh, should have a constant sum.

Examples:

1. A roller coaster starts on top of a 60 meter hill traveling at 5 meters per second. It goes to the bottom of a hill that is 20 feet off the ground. How fast is it going at the bottom of the hill?

The total energy, its kinetic and potential, at the top of the hill must be the same as at the bottom of the hill. Thus:

$KE_1 + PE_1 = KE_2 + PE_2$

$$\frac{1}{2}mv_1^2 + mgh_1 = \frac{1}{2}mv_2^2 + mgh_2$$

Each side can be divided by "m" and then values substituted to get:

$$\frac{1}{2} \cdot 5^2 + 9.8 \cdot 60 = \frac{1}{2}v_2^2 + 9.8 \cdot 20$$

Solving this equation finds that the roller coaster is going 28.44 meters per second or 63.6 miles per hour.

Gravitation

Isaac Newton found that gravity is a function of the masses of the objects and the distance between them. He also found that there is a universal gravitational constant of $G = 6.67 \times 10^{-11}$. An equation relating these variables is:
$F = \frac{Gm_1m_2}{r^2}$.

Example:
The force exerted between the Moon and Earth is 1.985×10^{26} newtons. The distance between them is 384,402 kilometers. If the mass of the Earth is 5.97×10^{24} kg, what is the mass of the Moon?

$$1.985 \times 10^{26} = \frac{(6.67 \times 10^{-11})(5.97 \times 10^{24})m_2}{(3.844 \times 10^5)^2}$$ [Answer: 7.37×10^{22} kilograms]

Kepler's Third Law

Johannes Kepler developed three laws of planetary motion. The first states that orbits around a body are elliptical in shape. The second states that equal areas within that ellipse are swept out in equal times. The third states that for any two orbiting bodies, the squares of the orbital periods (the amount of time it takes to complete a path around the orbited body) are directly proportional to the cubes of the semi-major axes of the orbits.[10] For our situation, the semi-major axis can be thought of as the mean distance from the Earth to the Sun. It can be found by averaging Earth's closest and farthest distance from the Sun.

Examples:
1. The mean distance from the Earth to the Sun is 93.0 million miles. Mars' mean distance is 141.6 million miles. Using Kepler's third law, if it takes the Earth a year to complete its orbit, how long should it take Mars?

$$\frac{93^3}{1^2} = \frac{141.6^3}{x^2} \quad \rightarrow \quad 93^3x^2 = 141.6^3 \quad \rightarrow \quad x = 1.88 \text{ Earth years}$$

2. Mercury orbits the Sun in 0.241 Earth years. What is Mercury's mean distance from the Sun?

$$\frac{93^3}{1^2} = \frac{x^3}{0.241^2} \quad \rightarrow \quad x = 36.0 \text{ million kilometers}$$

Above, it was seen that in comparing two orbiting bodies, the ratio $\frac{T^2}{r^3}$, is a constant. This ratio can be used in the following equation[21]: $\frac{T^2}{r^3} = \frac{4\pi^2}{G(m_1 + m_2)}$.

T is the number of seconds in a complete orbit
r is the distance between the bodies in meters
G is Newton's gravitational constant 6.6726×10^{-11}
m_1 and m_2 are the masses of the bodies in kilograms

Using this formula is a challenge. Likely, units will have to be converted and scientific notation will need to be used, but it does present a good application.

Examples:
1. The Earth is 149,600,000 kilometers from the Sun and it takes a year to complete its orbit. What is the mass of the Sun?
Because of the massive difference in size, the weight of the Earth can be disregarded without changing the answer very much. Making necessary conversions and substituting values gives:

$$\frac{(3.156 \times 10^7)^2}{(1.4957 \times 10^{11})^3} = \frac{4\pi^2}{(6.6726 \times 10^{-11})(m)} \quad \rightarrow \quad m = 1.989 \times 10^{30} \text{ kilograms}$$

2. It takes the Moon 27.2 days to orbit the Earth. The masses of the Moon and Earth are 7.35×10^{22} and 5.9737×10^{24} kilograms, respectively. How far apart are they?

$$\frac{(2.35 \times 10^6)^2}{x^3} = \frac{4\pi^2}{(6.6726 \times 10^{-11})(6.0472 \times 10^{24})} \quad \rightarrow \quad x = 3.83 \times 10^8 \text{ meters}$$

$E = mc^2$

Albert Einstein's famous energy equation, $E = mc^2$, shows that even a small amount of matter can be converted into a great deal of energy. For this equation, E is the energy in joules, m is mass in kilograms, and c is 3×10^8 m/s, the speed of light.

Examples:
1. There was an estimated 6.3×10^{13} joules of energy released from the explosion of the first atomic bomb. How much mass does that correspond to?
$(6.3 \times 10^{13}) = m(3 \times 10^8)^2$ [Answer: $m = 7 \times 10^{-4}$ kg or about 0.002 pounds]
2. In the Sun, 1 kilogram (a little over two pounds) of hydrogen fuses to become 0.993 kilograms of helium, with 0.007 kilograms becoming energy.[11] How much energy is produced in this process?
$E = (7 \times 10^{-3})(3 \times 10^8)^2$ [Answer: $E = 6.3 \times 10^{14}$ joules or ten times the first A-bomb]

Inflation

The formula $A = P(1 + r)^t$ could apply to growth in many different situations (inflation, savings accounts, the gross national product) that might be increasing at a certain annual percentage. At times, these problems may be, or perhaps need to be, solved with logarithms. However, many times that is not necessary.

Examples:

1. If inflation is at 2 percent annually, how much would a $520 television cost in three and a half years?

$$A = 520(1.02)^{3.5} = 557.32$$

2. What is the rate of inflation if a $41 costs $47 four years later?

$$47 = 41(1 + x)^4 \quad \rightarrow \quad 1.146 = (1 + x)^4 \quad \rightarrow \quad \sqrt[4]{1.146} = x \quad \rightarrow \quad x = 1.0347$$

The annual rate of inflation would be 3.47 percent.

Electrical Equations

There are important equations in electronics that show the relationship between important electrical concepts of power (in watts), voltage (volts), current (amperes), and resistance (ohms): $P = \dfrac{V^2}{R}$ and $P = I^2R$.

Examples:

1. A 12 volt battery supplies a bulb whose resistance is 18 ohms. What is the power in the circuit?

$$P = \frac{V^2}{R} = \frac{12^2}{18} = 8 \text{ watts}$$

2. A resistance of 80 ohms in a circuit consumes 1,200 watts of power. What is the voltage? [Answer: 309.8 volts]

3. What is the current in the above circuit? [Answer: 3.9 amps]

Poiseuille's Law

Jean Poiseuille (pronounced "Pwahzee") was a French physicist born in Paris in 1797. Poiseuille's equation is a good example of an equation containing a higher power. It describes the velocity of a liquid flowing through a pipe. This equation holds for the plumbing pipes in our homes, IV tubing, or blood vessels.

The formula[12] is $Q = \dfrac{\Delta P \pi r^4}{8vl}$, where r and l are the radius and length of the pipe, ΔP measures the difference in pressure between the ends of the pipe, and v is a measure of the viscosity (thickness) of the liquid.

Body-Mass Index

Body-Mass Index (BMI) is used to help determine the amount of a person's body fat. It combines a person's weight and height and is used as an overall indicator of health. Although BMI was developed in the 1800s, it did not gain popularity in the medical community until the 1980s. Health care professionals consider a BMI between 18.5 and 24.9 to be in the "normal" category.

$BMI = \frac{703w}{h^2}$ where w is weight in pounds and h is height in inches.[13]

Examples:

1. The BMI of a 5 foot 7 inch, 157 pound individual would be:

$$BMI = \frac{703 \cdot 157}{67^2} = 24.59$$

2. What range of weights could this person have and still be in the "normal" range?

This can be answered by solving the compound inequality:

$$18.5 \leq \frac{703w}{67^2} \leq 24.9$$

[Answer: 118.1 to 159.0 pounds]

Atmospheric Pressure

There are formulas for the amount of pressure the atmosphere exerts. This varies by elevation. Also, because there are different layers within the atmosphere, different formulas are used to accurately measure the pressure. It gets a little complicated. For two of the levels, first one has to use a formula to compute temperature at a specific elevation. Then that value is used in another formula to find the pressure. The equations from NASA give mathematical models and are produced by averaging measurements in both time and space. For these formulas, T is the temperature in degrees Fahrenheit, h is the altitude in feet, and P is pressure in pounds per square inch.[14]

The troposphere goes from sea level to 36,152 feet.

$T = 59 - 0.00356h$
$P = 2116[(T + 459.7)/518.6]^{5.256}$

The lower stratosphere exists between 36,152 and 82,345 feet.

$T = -70$
$P = 473.1e^{(1.73-0.000048h)}$

The upper stratosphere consists of elevations above 82,345 feet.

$T = -205.05 + 0.00164h$
$P = 51.97[(T + 459.7)/(389.98)]^{-11.388}$

For an additional mathematical application, students could compose the functions. This could be done to find a single equation for pressure based on elevation. For example, the troposphere formulas could be rewritten replacing T with T(h) and P with P(T), to find P(T(h)).

$$P(T(h)) = 51.97[(-205.005 + 0.00164h + 459.7)/389.98]^{-11.388}$$
$$= 51.97[(254.695 + 0.00164h)/389.98]^{-11.388}$$

Wind Chill Factor

Wind chill is a measure of how cold it feels to us. If the temperature is 33 degrees Fahrenheit, water won't freeze, regardless what the wind does. However, when the wind comes up, it will *feel* colder to us. When the wind picks up, heat is drawn from the body more quickly. According to the latest formula in use, a 30 degree temperature, with a wind of 10 miles per hour, has a wind chill of 21 degrees.

The concept of wind chill goes back to 1939. For the following 60 years, the formula for wind chill was[15]:

$$\text{Wind chill} = 0.0817(3.7\sqrt{v} + 5.81 - 0.25v)(t - 91.4) + 91.4t$$

Since 2001, an updated formula has been used. The National Weather Service currently uses[16]:

$$\text{Wind chill} = 35.74 + 0.6215T - 35.75v^{0.16} + 0.4275tv^{0.16}$$

For both equations, v is velocity in miles/hour and t is temperature in degrees Fahrenheit

Example:
Armed with a calculator, students could verify that a 30 degree temperature with a 10 mph wind does indeed correspond to a wind chill of 21 degrees. Students could then also try the old equation. It gives a wind chill temperature of 16 degrees.

Solving Radical Equations

Interest Rate over Two Years

If the price of an item goes up a certain percentage one year and another percentage the next, what is the equivalent annual increase over the two year period? If those percentages are written as decimals, x and y, that two-year increase can be found with the equation, $1 + P = \sqrt{(1 + x)(1 + y)}$.

Examples: An item has increased in cost 9 percent over two years. The first year it went up 7 percent. How much did it go up the following year?

$$1.09 = \sqrt{(1.07)(1 + y)} \quad \text{[Answer: x = 0.1104 or 11.04 percent]}$$

Surface Area of a Cone

There are a number of area and volume formulas that contain roots and radicals. One of these formulas is used to find the surface area of a cone with height h and radius r: $A = \pi r^2 + \pi r \sqrt{h^2 + r^2}$

Example:

A large funnel is to be constructed and will be used to transport grain from a silo into trucks. The base needs to be 8 feet across. There are 200 square feet of material to use to build the cone. What is the maximum height the cone could have?

$$200 = 4^2\pi + 4\pi\sqrt{h^2 + 16}$$ [Answer: h = 11.2 feet]

Coefficient of Restitution

When a basketball is dropped, it will bounce back up, but due to gravity, not quite as high as where it started. However, you would expect that basketball would get closer to its original height than a softball would. In science terms, it could be stated that the basketball has a higher coefficient of restitution than the softball. The Coefficient of Restitution = $\sqrt{\frac{h_2}{h_1}}$, where h_1 is the height from which the object is dropped and h_2 is the height to which it rebounds.[17]

Examples:

1. The NCAA has stated that official basketballs, when dropped from a height of 72 inches, must return to a height between 49 and 54 inches.[18] What is the coefficient of restitution for each of these heights? [Answer: 0.8250 and 0.8660]

2. A ball is known to have a coefficient of restitution of 0.7. If dropped from a height of 6 feet, to what height would it be expected the ball would rebound? [Answer: 2.94 feet]

$$\sqrt{\frac{h}{6}} = 0.7 \quad \rightarrow \quad 2.94 \text{ feet}$$

Pendulums

The period of a pendulum is the amount of time it takes to go through one complete back and forth cycle. Using metric system units, the formula is $T = 2\pi\sqrt{\frac{l}{g}}$, where T = the period in seconds, l is the length in meters, and g is the acceleration due to gravity, which is 9.8 m/sec².

The period depends only on the length of the pendulum. A boy is in a swing that takes 2.3 seconds to go back and forth. As gravity does its work, the boy's swing will go slower and won't go as high, but it will still take 2.3 seconds to complete a cycle.

Grandfather clocks use a pendulum to keep time. The first grandfather clock was developed in 1656. They are so tall because the pendulums have to be a certain length to keep the correct time. Grandfather clocks typically are constructed to have a period of 2 seconds. Smaller pendulum clocks could be constructed that have periods of one or one-half seconds. Grandfather clocks do have to be wound, but that is only to overcome the effects of gravity.

Example:

For the grandfather clock whose "tick-tock" is 2 seconds, the length of the pendulum would be found with the equation:

$$2 = 2\pi\sqrt{\frac{1}{9.8}} \quad \rightarrow \quad 1 = 0.993 \text{ meters}$$

The length necessary to have a pendulum clock that has a period of one-half or one second could similarly be found.

Speed vs. Skid Marks

Police can measure the length of skid marks on a road to determine the approximate speed a car was going from the time when the brakes were applied to when the car came to a stop. Obviously, longer skid marks would generally correspond to a greater speed. Those estimates of speed can be obtained from equations such as $S = \sqrt{30df}$, where S is the speed of the car in miles per hour, d is the length of the skid mark in feet, and f is the drag factor of the road surface.[19] This equation is assuming well-working brakes and that the car skids to a complete stop. The drag factor is a number between zero and one, and is a measure of how the road surface affects the tires of the car. Asphalt might typically have a drag factor of 0.75, while gravel might be 0.5, and for a snow covered road it might be 0.3.

Examples:

1. If a car left skid marks of 60 feet on an asphalt roadway, what is the car's approximate speed before braking?

$$S = \sqrt{30(0.75)(60)} \quad \rightarrow \quad S = 36.7 \text{ miles per hour}$$

2. To determine the skid factor for a road, a test was done. A car going 50 mph left skid marks of 100 feet. Find the value of f for this surface. [Answer: f = 0.83]

Speed of a Car Around a Curve

There is a road sign recommending an upcoming curve be taken at 30 miles per hour. How might that 30 mph value be determined in the first place? The maximum speed can be determined from the formula $V = \sqrt{rg\mu}$, where r is the radius in feet of a circle of which this curve is an arc, g is the acceleration due to gravity (32 feet/sec²), and μ is the coefficient of friction.[20] The coefficient of friction is closely related to the drag factor we encountered when looking at skid marks. However, the drag factor can change depending on the slope of the road surface, while the coefficient of friction does not. Since we'll deal with unbanked turns here, the values will be the same as before. So again, typical values would be asphalt 0.75, gravel 0.5, and snow 0.3.

Examples:
1. How fast should a car go around a snow covered curve of radius 200 feet? [Answer: 44 miles per hour]
2. On an asphalt road with a posted speed of 55 miles per hour, what should be the expected radius of the turn? [Answer: 126 feet]

While trigonometry applications are dealt with later in the book, what of the case of the car going around a banked turn?

A turn that is banked at θ degrees should have a maximum velocity of[21]:

$$V = \sqrt{\frac{rg(\sin\theta + \mu\cos\theta)}{\cos\theta - \mu\sin\theta}}$$

It can be seen that in the case of an unbanked turn (θ = 0°), this equation does simplify to our original equation $V = \sqrt{rg\mu}$.

Distance to the Horizon

No matter how good a person's eyesight is, one can only see so far due to the curvature of the Earth. Assuming a perfect sphere with no hills blocking the view, a good estimate of how far a person can see is $d = \sqrt{1.5e}$, with e being the elevation of the observer in feet and d is the distance that can be seen in miles.

While the equation itself is a good application, its derivation is not too difficult and uses a number of pieces of mathematics.

Taking the radius of the Earth, r, and, for the time being, calling our elevation h, how far is the distance, d, we can see? All of these distances are measured in miles. The line of sight to the horizon is tangent to the sphere. From geometry it is known that, at the point of tan-

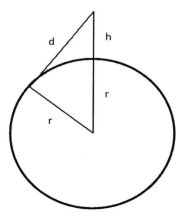

Looking to the horizon.

gency, the tangent line and the radius form a right angle. So the Pythagorean Theorem can be used.

$$d^2 + r^2 = (r + h)^2 \quad \rightarrow \quad d^2 + r^2 = r^2 + 2rh + h^2$$
$$\rightarrow \quad d^2 = r^2 + 2rh + h^2 - r^2$$
$$\rightarrow \quad d^2 = 2rh + h^2$$
$$\rightarrow \quad d = \sqrt{2rh + h^2}$$

This is a perfectly good formula, but a simpler one can be obtained by taking a few liberties. The h^2 term doesn't really add much value to the formula. Leave that term out and substitute the Earth radius of 3,960 miles. While we're at it, the elevation is probably more conveniently measured in feet. If e is the elevation in feet, $h = \frac{e}{5,280}$. Letting $h^2 = 0$, $r = 3,960$, and simplifying gives our formula: $d = \sqrt{2rh} = \sqrt{2(3,960)\frac{e}{5,280}} = \sqrt{1.5e}$.

Examples:

1. A person is standing at the lake shore. Her eyes are 5.5 feet above lake level. How far across the lake can she see? [Answer: 2.9 miles]

2. A lighthouse is to be constructed so its light can be seen 10 miles out to sea. How many feet above sea level should the light be located? [Answer: 66.7 feet]

3. How far can you see from the top of the observation deck of the Empire State Building (elevation 1,210 feet)? [Answer: 42.6 miles]

Level of Confidence

A survey states that 52 percent would vote for Millard Fillmore for president. Often surveys might also state that there is a 3 percent margin of error with a 95 percent level of confidence. That is to say that the surveyors are 95 percent certain that if you were to ask every voter, the actual result is somewhere between 49 percent and 55 percent.

The U.S. Constitution says that a census must be conducted every ten years. It is a huge undertaking that takes months to accomplish. Then how are album sales, television ratings, and unemployment statistics published monthly or even weekly? Those results are based on surveys. The only way to get to a 100 percent level of confidence is to ask everyone, which isn't realistic. The number of people that would need to be surveyed is dependent upon the level of confidence and margin of error desired. It turns out the number of people in the total population isn't a factor. Even though millions of people watch television, the number of families surveyed is relatively small in comparison to the nation as a whole. The Nielsen ratings by 2017 plan to increase their number of families, but that will still only be 6,200 for any given week.[22]

The following formulas find the number that need to be surveyed, n, for a particular level of confidence. (The U.S. Census Bureau uses a 90 percent level of confidence for its unemployment numbers.[23])

99 percent level of confidence: $\frac{1.29}{\sqrt{n}}$

95 percent level of confidence: $\frac{0.98}{\sqrt{n}}$

90 percent level of confidence: $\frac{0.82}{\sqrt{n}}$

Examples:

1. Two thousand people are surveyed. If there is a desired 99 percent level of confidence, that would give what percent confidence interval?

$$\frac{1.29}{\sqrt{2000}} = \pm 2.9\%$$

2. For a 95 percent level of confidence, how many people would have to be surveyed to have a confidence interval of 4 percent?

$4\% = 0.04 = \frac{0.98}{\sqrt{n}}$ $\rightarrow$ $n = 600.25$ $\rightarrow$ So, at least 601 people would need to be surveyed.

Escape Velocity

Could you jump so that you could go into outer space? Unlikely, but if you could jump with enough speed, it could happen. Escape velocity is the velocity it takes an object to leave a body's gravitational pull. The formula is:

$V = \sqrt{\frac{2Gm}{r}}$, where V = escape velocity in meters per second

$G = 6.67 \times 10^{-11}$

m = mass of planet/star/moon in kilograms,

r = radius of planet/star/moon in meters

Examples:

1. The Moon has a radius of 1,737.5 kilometers and a mass of 7.35×10^{22} kilograms. What is the Moon's escape velocity?

$V = \sqrt{\frac{2(6.67 \times 10^{-11})(7.35 \times 10^{22})}{1.7375 \times 10^{6}}} = 2,376$ m/sec (5,315 miles per hour)

2. Pluto has an escape velocity of 1.2 km/sec and has a radius of 1.15×10^{6} meters. What is the mass of Pluto? [Answer: 1.24×10^{22} kg]

3. A body that is extremely small and heavy would have an incredibly strong gravitational force. It would, thus, have a very large value for its escape velocity. What happens if that escape velocity was even greater than the speed of light (3×10^{8} m/s)? Then nothing, not even light, could escape. If light can't get out, that body is for all practical purposes, invisible, or a "black hole." What would the radius of the Earth (a mass of 5.972×10^{24} kg) have to be to be a black hole?

$$3\text{x}10^8 = \sqrt{\frac{2(6.67\text{x}10^{-11})(5.972\text{x}10^{12})}{r}} \quad \rightarrow \quad 9\text{x}10^{16} = \frac{79.66648\text{x}10^{13}}{r}$$

$$\rightarrow r = 0.0885 \text{ meters}$$

A dying star could collapse in on itself with such force that it shrinks and becomes a black hole. For this to happen to the Earth, it would have to collapse to roughly the size of a marble.

Free Falling Objects

We previously looked at dropping objects. Here are another couple of equations to describe this type of motion. If an object is dropped, with no initial velocity, the formula for how fast it is going and how long it is falling can be found with the formulas: $v = \sqrt{2gh}$ and $t = \sqrt{\frac{2h}{g}}$, where g is 9.8 m/s², h is the distance fallen (m), t is time (sec), and v is its velocity (m/s).

Examples:

1. An object is dropped from 100 meters. How fast is it going when it hits the ground? [Answer: 44.27 m/s]

2. An object is dropped and is going 50 m/s when it lands. From what height was it dropped? How many seconds has it been falling? [Answers: 127.55 m, 5.1 seconds]

3. Suppose, on another planet, this same object only drops 70 meters in 5.1 seconds. What is the value for g on this planet and how fast is it dropping when it lands? [Answers: g = 5.383 m/s², 27.45 m/s]

Geostationary Orbits

A satellite is orbiting Earth. It must maintain a certain velocity for a given altitude in order to stay in orbit. That velocity is $V = \sqrt{\frac{Gm}{r}}$. The time it takes to complete this orbit is found with the formula $T = 2\pi\sqrt{\frac{r^3}{Gm}}$. For these equations,

V—velocity in meters/second
T—the period of the orbit, in seconds.
r—meters from the center of the object being orbited, in this case Earth.
G—the gravitational constant 6.67×10^{-11}
m—the mass of the orbited object

A lot of information could be gained from this formula. An interesting point is what happens if the period of the satellite happened to be 24 hours, the same as the Earth's. That satellite would always be in the same location above the Earth. It turns out this is often a desired effect for satellites. Staying in the same location above the Earth is important in relaying telephone calls or television programs. This is also the basis of the global positioning system. When homes in North America point their satellite dish to the south, it is because it is pointed at a satellite orbiting above the Equator. To find the desired orbital height, the above formula would be used, substituting 86,400 (number of seconds in a day) for T, and 5.98×10^{24}kg (the mass of the Earth), and solving for r. Having found r, the velocity of the orbiting satellite could be found using the other equation.

Ponderal Index

Like BMI, the ponderal index (PI) is a measurement used to indicate a healthy height/weight amount. The ponderal index is most commonly used in assessing newborn children. A healthy ponderal index is considered to be between 12.49 and 13.92. In English units, the formula is as follows[24]:

$PI = \dfrac{h}{\sqrt[3]{w}}$, where h is the height in inches and w is the weight in pounds.

Example:
What would a 26-inch child weigh to be in the normal range?

$12.49 \leq \dfrac{26}{\sqrt[3]{w}} \leq 13.92$ [Answer: Between 6.5 and 9.0 pounds]

Theory of Relativity

The idea that time, length, and mass are relative quantities did not originate with Albert Einstein. Hendrik Lorentz had developed transformation equations that quantified the changes. Einstein, however, gave the correct interpretation to why the changes take place.[25] The Lorentz transformation equations compared length, mass, and time for bodies at rest to bodies in motion relative to the observer. The observed changes in the resting length, mass, and time; l_0, m_0, and t_0; to that of those bodies in motion relative to the observer are found as follows:

$$ l = l_0 \sqrt{1 - \frac{v^2}{c^2}}, \qquad m = \frac{m_0}{\sqrt{1 - \frac{v^2}{c^2}}}, \qquad \text{and } t = \frac{t_0}{\sqrt{1 - \frac{v^2}{c^2}}} \text{ where v is the object's} $$

velocity and c is the speed of light.

Examples:
1. A rocket traveling at 60 percent the speed of light flies over a 100 yard football field. How long will the field appear to the pilot?

$$ l = 100 \sqrt{1 - \frac{(0.6c)^2}{c^2}} \;=\; 100\sqrt{1 - 0.36} \;=\; 80 \text{ yards} $$

2. The length of a meter stick is seen by an observer as being 0.7 meters. How fast is that observer traveling?

$$0.7 = 1.0\sqrt{1 - \frac{v^2}{c^2}} \quad \rightarrow v = 0.714c \text{ or traveling at } 71.4 \text{ percent of the speed}$$

of light.

3. A 300-kilogram astronaut is in flight traveling at 24,791 miles per hour (0.004 percent the speed of light). What is the astronaut's mass during the flight?

$$t = \frac{300}{\sqrt{1 - \frac{(.00004c)^2}{c^2}}} = 300.0000002 \text{ kilograms}$$

The fastest any human has traveled was a speed of 24,791 miles per hour on the Apollo 10 mission to the moon. This is approximately 0.004 percent of the speed of light.[26] Just as the astronauts on that flight will probably not notice the 0.0000002 kg weight gain, most effects of relativity aren't readily apparent to us. However, some technology such as the global positioning system could not function without taking the effects of relativity into account. The CERN particle collider has accelerated particles to 99.99 percent of the speed of light with results consistent with relativity's predictions.

Einstein stated that nothing can travel faster than the speed of light. Mathematically speaking, students could substitute c or 1.1c for velocity in any of the equations and consider how those answers might be interpreted.

Solving Rational Equations

Simplifying Complex Fractions

Complex fractions are fractional expressions within a fraction. They can be simplified into a more convenient form.

Examples:

1. Inflation rose 0.3 percent last month. What is this rate in fraction form?

$$\frac{\frac{3}{10}}{100} = \frac{\frac{3}{10}}{100} \cdot \frac{10}{10} = \frac{3}{1000}$$

2. You have completed one and a quarter miles of a three and a half mile race. What fraction of the race have you finished?

$$\frac{1\frac{1}{4}}{3\frac{1}{2}} = \frac{\frac{5}{4}}{\frac{7}{2}} = \frac{\frac{5}{4}}{\frac{7}{2}} \cdot \frac{4}{4} = \frac{5}{14}$$

Averaging Rates

The formula for the harmonic mean of two numbers can be used to average rates: $M = \dfrac{2}{\frac{1}{a} + \frac{1}{b}}$.

Example:

1. Going on a round trip, the speed going in one direction is 50 mph and coming back is 40 mph, what is the speed for the entire trip? The intuitively expected answer of 45 is not quite correct, as more time is spent going at the slower rate.

$$R = \frac{2}{\frac{1}{50} + \frac{1}{40}} = 44.4 \text{ mph}$$

2. On a three-day trip, you are hoping for an average speed of 40 miles per hour. The first day you average 60 and the next day 45 mph. What minimum speed do you have to average on the third day to reach your goal of 40 mph?

$$40 = \frac{3}{\frac{1}{60} + \frac{1}{45} + \frac{1}{x}} \quad \rightarrow \quad 40 = \frac{3}{\frac{1}{60} + \frac{1}{45} + \frac{1}{x}} \cdot \frac{180x}{180x} \quad \rightarrow \quad 40 = \frac{540x}{7x + 180} \quad \rightarrow \quad x = 27.7 \text{ mph}$$

3. The same is true in averaging miles per gallon when the distance traveled is the same. Again, even though the distance covered is the same, there is more time spent at the lower fuel mileage. So, the mileage is a little less than if finding the mean average of the numbers. A car averages 20 mpg one way and 30 mpg on the way back. The average miles per gallon for the entire trip is:

$$x = \frac{2}{\frac{1}{20} + \frac{1}{30}} \cdot \frac{60}{60} = \frac{120}{3 + 2} = 24 \text{ miles per gallon}$$

4. If a stock is trading at $24 per share and the earnings over the past year was $1.25 per share, that stock would have a Price to Earnings, or P/E, ratio of 19.2. An average P/E ratio of several stocks is found by finding the harmonic mean of those stocks. If two stocks have P/E ratios of 19.2 and 18.6, what would the P/E ratio of a third stock have to be to have an overall ratio of 20.0?

$$20 = \frac{3}{\frac{1}{19.2} + \frac{1}{18.6} + \frac{1}{x}} \quad \rightarrow \quad 20 = \frac{3(19.2)(18.6)x}{37.8x + 357.12} \quad \rightarrow \quad x = 22.65$$

Power-Speed Numbers

In baseball statistics, the power speed number[27] ranks players highly that are both strong and fast. It combines their home runs (h) and their stolen bases (s) by finding their harmonic mean: Power-Speed Number $= \frac{2}{\frac{1}{h} + \frac{1}{s}}$ Some sources will write this expression as the equivalent $\frac{2hs}{h + s}$ This can be shown to be equivalent by simplifying the complex fraction.

Example:

In 1956, Willie Mays hit 36 home runs and had a power speed number of 37.9. How many stolen bases did he have that year?

$$37.9 = \frac{2 \cdot 36s}{36 + s} \quad \rightarrow \quad (36 + s)37.9 = 72s \quad \rightarrow \quad s = 40.01, \text{ thus 40 stolen bases.}$$

More Work Problems

Earlier, we looked at work problems that resulted in quadratic equations. This section will focus on work problems that contain fractions and are first degree equations.

If it is known how fast a person or machine is working on a task, it is easy to tell how long it will take to finish. If there are multiple parties at work on the same project, it becomes just slightly harder.

Examples:

1. Working on its own, a particular pump can empty a pool in 6 hours. If a second pump is set up that could empty the pool in 5 hours, how long would the job take with both of the pumps working at the same time?

The first pump can do one-sixth of the job in an hour. The fraction $\frac{x}{6}$ represents the amount that can be done in x hours. Similarly the second pump can accomplish $\frac{x}{5}$ of the job. To do one whole job is represented by the equation:

$$\frac{x}{6} + \frac{x}{5} = 1$$

Multiplying each side by 30 gives the equation $6x + 5x = 30$. Solving for x finds it would take about 2.7 hours to drain the pool.

2. You would like to be able to drain your pool in two hours. How fast would the second pump have to be for this to happen?

You would need to solve the equation: $\frac{1}{6} + \frac{1}{x} = \frac{1}{2}$. Solving the equation gives an answer of $x = 3$. So, if working alone, this pump would have to be one that could drain the pool in three hours.

Focal Length

Every lens has a particular focal length. Light rays go through the lens and converge at a single point. This distance from the lens to this focal point is the focal length (f). An object that is d_o units from the lens will have an image d_i units from the lens. These variables are related by the equation[52]:

$$\frac{1}{f} = \frac{1}{d_o} + \frac{1}{d_i}$$

Light from an object going through a lens will form an image. By measuring those distances and using this formula, the focal length can be found.

Examples:

1. An object that is 25 cm away from a lens has an image 10 cm from the lens. What is the lens' focal length? [Answer: 7.1 feet]

2. An object is 10 inches away from a lens that has a focal length of 2 inches. At what distance will the image be formed?

Substituting gives the equation: $\frac{1}{2} = \frac{1}{10} + \frac{1}{x}$. Solving the equation shows that the image is 2.5 feet from the lens.

How does a photographer make sure a picture is in focus? The photographer makes manual adjustments to the camera which brings the picture into focus. When doing so, he is actually moving the lens slightly forward or backward inside the camera. He is changing the object distance and image distance, until the picture looks right for the focal length for that particular lens. Autofocus works similarly. An infrared beam of light can find the distance to the object. Knowing that distance, the camera then shifts the lens the appropriate amount.[28]

A lens might state it is an 18–135 mm zoom lens. For this type of lens, the focal length actually does change—in this case, anywhere from 18 mm to 135 mm. Inside a zoom lens are a number of elements that are moved, changing the effective focal length as a whole. Much more impressive than any camera is the eye. If an image is not in focus, muscles surrounding the eye flex or relax changing the shape of the eye, which changes the focal length. This happens countless times during the day without our awareness.

The Lensmaker's Equation

For a given lens, the lensmaker's equation relates the focal length and the curvature of the lens.[29]

$$\frac{1}{f} = (n-1)\left(\frac{1}{r_1} - \frac{1}{r_2}\right)$$

Each lens has two sides. Typically, each side is the arc of a circle. One side of the lens has a radius of curvature of r_1 and the other side of the lens has a radius of curvature of r_2. The index of refraction for the lens is n. (For example, in a vacuum, n = 1; in air, n = 1.0003; and in water, n = 1.33) The focal length of the lens is again f.

Examples:

1. A lens has values of r_1 = 20 cm, r_2 = –10 cm (it is written as a negative because it is curved in the other direction), and for our lens, n = 1.5. What is the focal length of the lens?

$$\frac{1}{f} = (1.5 - 1)\left(\frac{1}{20} - \frac{1}{-10}\right) \quad \rightarrow \quad \frac{1}{f} = \left(\frac{1}{40} + \frac{1}{20}\right) \quad \rightarrow \quad f = 13.33 \text{ cm.}$$

2. Suppose we need lenses made for glasses that will correct for near-sightedness. To accomplish this, r_1 needs to be greater than r_2. Both are in the same direction, so both values are positive. Suppose the focal length is -80 (the focal length is a negative value if it acts as a divergent lens), $n = 1.5$, and $r_1 = 40$. Find the value of r_2. [Answer: $r_2 = 20$]

IV Flow Rate

Patients in a hospital are often put on an intravenous, or IV, solution. An IV can be set manually for a certain flow rate in drops per minute. A formula used to determine the flow rate is based on volume (mL), time (minutes), and a drop factor (drops per mL)[30]:

$$\text{Flow Rate} = \frac{\text{volume}}{\text{time}} \cdot \text{drop factor}$$

Examples:

1. How long should an IV be administered for 900 milliliters of medicine that is to have a flow rate of 150 drops per minute and a drop factor of 20 drops per milliliter?

$$150 = \frac{900}{x} \cdot 20 \quad \text{[Answer: 120 minutes]}$$

2. A medicine needs to be administered at a flow rate of 28 drops per minute for one hour with a drop factor of 40 drops per milliliter. What volume of medicine is needed?

$$28 = \frac{x}{60} \cdot 40 \quad \text{[Answer: 42 milliliters]}$$

Dosage Amounts

A normal dose of medicine for a child is usually different than that for an adult. There are formulas that can be used to estimate the proper dosage. One is based on the weight and another is based on the age of the child. In the following two formulas, C is the child's dose, D is the adult dose, W is the child's weight in pounds, and A is the age of the child.[31]

$$\text{Clark's Rule: } C = D \cdot \frac{W}{150} \qquad \text{Young's Rule: } C = D \cdot \frac{A}{A + 12}$$

Example:

A particular medicine prescribes a daily maximum amount of 8 fluid ounces for an adult. This medicine is to be given to an 8-year-old, 70-pound child. Find appropriate amounts by using both Clark's and Young's Rules. [Answers: By Clark's Rule 3.7 ounces and by Young's rule 3.2 ounces]

1. An adult dose is 4 fluid ounces. Using Young's Rule, at what age would a dose of 1 ounce be appropriate?

$$1 = 4 \cdot \frac{A}{A+12} \quad \rightarrow \quad A+12 = 4A \quad \rightarrow \quad A = 4 \text{ years old}$$

2. An adult dose is 700 milligrams. Using Clark's Rule, what would be an appropriate child's weight for a dose of 300 milligrams?

$$300 = 700 \cdot \frac{W}{150} \quad \rightarrow \quad W = 64.3 \text{ pounds}$$

Radar Guns

The Doppler effect is used by police to detect a car's speed. If a car is stationary, a radar gun's waves should bounce off the car and return to the police officer at the same frequency. If the car is moving toward the officer, the waves will bounce off and return at a higher frequency. The higher the frequency, the faster the car must be going.

The equation $\Delta f = f\left(\frac{2v}{c}\right)$ can be used to detect the car's speed.[32] The frequency emitted from the radar gun is f, Δf is the change in the frequency, v is the speed of the car in m/sec, and c is the speed of light/radar (3×10^8 m/sec).

Example:
In a 50 mph zone, suppose the radar gun emitted waves with a frequency of 10.5×10^9 hertz. The waves coming back had increased by 1,500 hertz. Is the car going beyond the speed limit?

$$\Delta f = f\left(\frac{2v}{c}\right) \quad \rightarrow \quad 1,500 = 10.5 \times 10^9 \left(\frac{2v}{3 \times 10^8}\right) \quad \rightarrow \quad v = 21.43 \text{ m/sec}$$

This translates to just under 48 miles per hour. Carry on, citizen.

Electrical Resistance, Inductance and Capacitance

Rational equations are often used in electronics. One such equation is for calculating resistance (measured in ohms). Resistors are important in electronic circuits. They impede the flow of current. Resistance converts electrical energy into heat. In many cases, as little resistance as possible is wanted. However, resistance is desired in items where generating heat is important; such as heaters, toasters, and dryers.

Several resistors may be used in a circuit. Suppose there are three resistors in a circuit with values R_1, R_2, and R_3. If connected in series (one after the other) the total resistance in the circuit is simply the sum of the resistances: $R_T = R_1 + R_2 + R_3$. Three resistors connected in parallel have a total resistance of $R_T = \frac{1}{\frac{1}{R_1} + \frac{1}{R_2} + \frac{1}{R_3}}$.

Examples:
1. Two resistors of 20 and 40 ohms are connected in parallel. What would the resistance of a third resistor have to be to have a total resistance of 10 ohms?

$$10 = \frac{1}{\frac{1}{20} + \frac{1}{40} + \frac{1}{x}}$$

Multiplying the numerator and denominator of the fraction by 40x, gives 10 $= \frac{40x}{2x + x + 40}$ which will produce a value of x = 40. So, a third resistor of 40 ohms is needed.

2. A circuit has two resistors connected in parallel. If the total resistance is 12 ohms and one of the resistors is known to be 30 ohms, what is the value of the other resistor? [Answer: 20 ohms]

Inductors and capacitors act something like batteries, storing electrical energy. Inductance is measured in henrys and stores energy in magnetic fields. The formulas for inductance look very much as those for resistance.

In series: $L_T = L_1 + L_2 + L_3$ In parallel: $L_T = \frac{1}{\frac{1}{L_1} + \frac{1}{L_2} + \frac{1}{L_3}}$

Example:

1. Two inductors connected in parallel have a total inductance of 75 millihenrys. If one of the inductors is 300 millihenrys, what is the value of the second? [Answer: 100 millihenrys]

Capacitors are measured in farads and store energy in electric fields. The capacitance formulas are reversed from the pattern we've seen previously.

In series: $C_T = \frac{1}{\frac{1}{C_1} + \frac{1}{C_2} + \frac{1}{C_3}}$ In parallel: $C_T = C_1 + C_2 + C_3$

Example:

Two capacitors of 10 and 15 farads are connected in series. What is the total capacitance? [Answer: 6 farads]

Coulomb's Law

There is force acting between any two charged particles. We see this in action with static electricity, lightning, or the flow of electricity in a circuit. The amount of force between any two charged particles is dependent upon how strong the charges are and the distance between them. This formula was discovered by French scientist Charles Coulomb (pronounced "Koolohm") in the 1700s.

Coulomb's Law: $F = k\frac{q_1 q_2}{r^2}$

F is the force in newtons, the charges are measured in coulombs, r is the distance between the charges in meters, and k is a constant, 9×10^9.

This equation is similar in form to Isaac Newton's formula for gravity. As forces go, gravity is fairly weak and only has much effect on very large bodies. Coulomb's law deals with very small particles. Here is an example in which the denominator is unknown.

Example:
Suppose that between two particles with charges of 0.5 and 1 coulomb there is a force of 1.8×10^{11} newtons. What is the distance between these particles?

$$F = k\frac{q_1 q_2}{r^2} \quad \rightarrow \quad 1.8 \times 10^{11} = 9 \times 10^9 \frac{0.5 \cdot 1}{r^2} \quad \rightarrow \quad 0.02 = \frac{0.5}{r^2} \quad \rightarrow \quad r = 0.16 \text{ m}$$

Determinants

Area of a Triangle

The area of a triangle on a coordinate plane can be found using a 3x3 determinant. Take the three ordered pairs representing the three vertices of a triangle. A determinant can be formed in which the first column contains the x-values and the second column contains the y-values. The third column always contains three ones. One-half the value of this determinant is found. The absolute value is then taken, giving the final answer.

Example:
This works for any triangle, but let's take one we know the answer to ahead of time. The triangle with vertices (2,2), (2,7), and (5,2) is just half of a rectangle that has a length of 5 and a width of 3. So we know ahead of time that the triangle should have an area of 7.5.

$$0.5 \cdot \begin{vmatrix} 2 & 2 & 1 \\ 2 & 5 & 1 \\ 7 & 2 & 1 \end{vmatrix} = 0.5\left(2\begin{vmatrix} 5 & 1 \\ 2 & 1 \end{vmatrix} - 2\begin{vmatrix} 2 & 1 \\ 7 & 1 \end{vmatrix} + 1\begin{vmatrix} 2 & 5 \\ 7 & 2 \end{vmatrix}\right)$$

$$= 0.5(2(3) - 2(-5) + 1(-31)) = -7.5$$

The area is $|-7.5| = 7.5$

Another use for this technique is to determine if three points are collinear. If the points are collinear, a triangle cannot be formed, resulting in an area of zero.

Volume of a Tetrahedron

A tetrahedron is a pyramid shape made up of four triangles. Given the ordered triples that are the four vertices, there is a formula that is very similar to the one used to find the area of a triangle. The determinate has four rows that contain the four ordered triples. The last column is again all ones. This time the determinant is multiplied by one-sixth, and then the absolute value is taken.

Inverses of Matrices

Technically, matrix division isn't an operation. However, matrix multiplication can be reversed by multiplying by the inverse. Part of the process of finding the inverse of a matrix involves evaluating a determinant. Additionally, finding the determinant tell us if an inverse even exists in the first place. If the determinant is equal to zero, that matrix does not have an inverse.

Cramer's Rule

Cramer's rule can be used to solve a system of equations. It uses the coefficients of the equations to form determinants. Those determinants are evaluated to find the solutions. There are several ways to solve a system of equations and each method has its advantages and disadvantages. An advantage of Cramer's Rule is that computers and calculators can be used to do the heavy lifting of evaluating those determinants. This would be especially true for systems that contain more than two or three equations. There is time involved in entering the coefficients into the device, but after that, the solution is found in a split second.

Understandably, most, if not all, of the systems examined in a high school math class contains two or three equations. In the real world, systems can have many more than that. Writing to a reporter in April of 1932, Charles Ellis, the chief engineer working on the construction of the Golden Gate Bridge stated, "Any engineer, however familiar he may be with the general theory will, before he goes far with the problem, have to learn how to do certain things for there is considerable original work to be done. Incidentally, there are 113 equations containing 39 unknown quantities to be solved in checking up the design after it has been completed."[33]

Vectors

A three by three determinant can be used to find the cross product of two vectors. Using the vectors (a,b,c) and (c,d,f), and the unit vectors, the cross product is found with the determinant:

$$\begin{vmatrix} \vec{i} & \vec{j} & \vec{k} \\ a & b & c \\ d & e & f \end{vmatrix}$$

Matrices

A matrix is a collection of numbers arranged in rows and columns. Because of this rather general definition, matrices can be used to represent all kinds of information.

Sports Standings

Take the following three matrices that represent the win-loss records of three seasons for college basketball teams at Duke, North Carolina, and Kentucky respectively. Matrix A represents the 2011–12 season, B is the 2012–13 season, and C is the 2013–14 season.

$$A = \begin{bmatrix} 27 & 7 \\ 32 & 6 \\ 38 & 2 \end{bmatrix} \qquad B = \begin{bmatrix} 30 & 6 \\ 25 & 11 \\ 21 & 12 \end{bmatrix} \qquad C = \begin{bmatrix} 26 & 9 \\ 24 & 10 \\ 29 & 11 \end{bmatrix}$$

Information in matrices can easily be manipulated using rules for matrix and scalar operations.

Examples:

1. What is the improvement in the records from the 2012–13 to the 13–14 seasons? [Answer: C–B]

2. What are the combined records for these three years? [Answer: A + B + C]

3. If the NCAA decided to increase the length of the 2013–14 season by 20 percent, what might have been the predicted team's records for those three years? [Answer: A + B + 1.2C]

Solving Systems of Equations

There are several ways to solve a system of equations. It was seen previously that Cramer's rule can be used. Another way is to use augmented matrices. What is the solution for the system consisting of the following?

$-2x - 2y - 3z = -17$
$-x + 3y + z = 13$
$-3x + 2y - 3z = 3$

This system can be written as an augmented matrix by copying the coefficients into a matrix form.

$$\begin{bmatrix} -2 & -2 & -3 & -17 \\ -1 & 3 & 1 & 13 \\ -3 & 2 & -3 & 3 \end{bmatrix}$$

In due time, multiplying rows by various values, and addition and subtraction of rows, leads to the following matrix.

$$\begin{bmatrix} 1 & 0 & 0 & 4 \\ 0 & 1 & 0 & 6 \\ 0 & 0 & 1 & -1 \end{bmatrix}$$

Reattaching the variables, $1x + 0y + 0z = 4$, or $x = 4$. Similarly, $y = 6$ and $z = -1$.

Matrices can solve systems in another way. Taking the same set of equations as above, they could be expressed as the matrix equation:

$$\begin{bmatrix} 2 & -2 & -3 \\ -1 & 3 & 1 \\ -3 & 2 & -3 \end{bmatrix} \cdot \begin{bmatrix} x \\ y \\ z \end{bmatrix} = \begin{bmatrix} -17 \\ 13 \\ 3 \end{bmatrix}$$

The values for the variables could be found by multiplying each side by the inverse matrix. The process in general would look something like this:

$$A \cdot x = B \rightarrow A^{-1} \cdot A \cdot x = A^{-1} \cdot B \rightarrow x = A^{-1} \cdot B$$

Predicting the Future

If it is a sunny one day, it will likely be sunny the next. A weather front may move into the area to change this, but weather conditions tend to stay roughly the same for a while. Suppose that historically for your area, if it is sunny one day, there is an 80 percent chance it will be sunny the next. That, of course, means there is a 20 percent chance it won't be sunny. On the other hand, if it is not sunny, suppose the next day has a 60 percent chance of not being sunny and a 40 percent of being sunny. This information can be summed up in a transition matrix:

$$W = \begin{bmatrix} 0.8 & 0.2 \\ 0.4 & 0.6 \end{bmatrix}$$

The two rows represent the chance of sunny or not sunny, respectively. The two columns are the chance of sunny or not sunny on the next day.

This matrix can be used to predict what could happen not just today, but two days from now. The way to find this is to use matrix multiplication.

$$W^2 = \begin{bmatrix} 0.8 & 0.2 \\ 0.4 & 0.6 \end{bmatrix} \begin{bmatrix} 0.8 & 0.2 \\ 0.4 & 0.6 \end{bmatrix} = \begin{bmatrix} 0.72 & 0.28 \\ 0.56 & 0.44 \end{bmatrix}$$

Thus, if it is raining on a given day, there is a 72 percent chance it will be raining two days from now.

What about three days in the future?

$$W^3 = \begin{bmatrix} 0.688 & 0.312 \\ 0.624 & 0.376 \end{bmatrix}$$

If it is raining today, there is a 68.8 percent chance of rain three days from now.

Another example:

Eight times a team that won the Super Bowl has also won the next year's Super Bowl. This has happened approximately 17 percent of the time. Any random team that didn't win a Super Bowl has had a 3 percent chance of winning the next year. Placing this information into a transition matrix would look like this:

$$S = \begin{bmatrix} 0.17 & 0.83 \\ 0.03 & 0.97 \end{bmatrix}$$

The matrix $S^2 = \begin{bmatrix} 0.054 & 0.946 \\ 0.034 & 0.966 \end{bmatrix}$. This would represent a prediction for what would happen two years in the future.

According to this matrix, the probability of winning one year and then winning two years later has fallen to 5.4 percent. It has actually happened 4.3 percent of the time. It has happened twice—the Dallas Cowboys in 1994 and 1996 and the New England Patriots in 2002 and 2004.

These transition matrices represent what are called Markov Chains. There are many examples that demonstrate this process.

1. Whether customers will switch cell phone companies. (If a customer is with Verizon, what is the chance she will be with that company five years from now?)

2. The generational movement of genetic traits.

3. The generational movement between lower, middle, and upper income.

4. The shifting of voters between the political parties.

Direct and Indirect Wins

Everyone knows that if one team beats another in a particular contest, it doesn't mean that team is necessarily the better of the two. Those same two teams could play the next day and have an opposite result. More information might be helpful in deciding which might be the better team. If Team A beat Team B and later Team B beat Team C, that would seem to imply that Team A could defeat Team C. This "indirect win" should count for something. The number of these indirect wins could be found with matrix multiplication.

This would probably be more informative for a larger league and more games, but for now, suppose that there is a three-team league with the following results thus far:

Team A beats Team B, Team A beats Team C, Team B beats Team C
Team A beats Team B (again), Team C beats Team A, Team C beats Team B

We can summarize these results in a matrix. Team A, B, and C will be the rows and how they did against Team A, B, and C will be the columns.

$$M = \begin{bmatrix} 0 & 2 & 1 \\ 0 & 0 & 1 \\ 1 & 1 & 0 \end{bmatrix}$$

Team A beat Team B twice. That is why there is a 2 in the first row, second column. These are the results of direct wins. The indirect wins could be done by finding M^2. Then what about the total of the direct and indirect wins against each team? This can be found with matrix addition of $M + M^2$. The total number of wins, direct or indirect, can be found by simply adding the entries in each row.

$$M^2 = \begin{bmatrix} 1 & 1 & 2 \\ 1 & 1 & 0 \\ 0 & 2 & 2 \end{bmatrix} \qquad M + M^2 = \begin{bmatrix} 0 & 2 & 1 \\ 0 & 0 & 1 \\ 1 & 1 & 0 \end{bmatrix} + \begin{bmatrix} 1 & 1 & 2 \\ 1 & 1 & 0 \\ 0 & 2 & 2 \end{bmatrix} = \begin{bmatrix} 1 & 3 & 3 \\ 1 & 1 & 1 \\ 1 & 3 & 2 \end{bmatrix}$$

By this method, the top team appears to be Team A with a total of 7 direct and indirect wins. Right on their tails with 6 wins is Team C. Team B has 3.

Airline Flights

The previous idea could be used in a number of different areas. One example would be airline flight patterns. If a traveler is flying from Chicago to L.A., she might be able to go there directly, or she might book a flight in which there is a layover in Dallas.

Suppose there are the cities of A, B, C, and D. City A has direct flights to B and D. B only flies to A and C. C flies to A, B, and D. D flies only to A. This information can be summarized in matrix form. The matrix, we'll call F, showing all flights with layovers can be found by finding F^2. The totals of the flights, direct or with a layover, are found with the matrix that is $F + F^2$.

Transformations

A point or group of points can be moved to a different location by the use of matrices. A collection of n ordered pairs could be written as a 2 by n matrix. A group of ordered triples could be written as a 3 by n matrix. Translations, rotations, reflections, and dilations can be performed by using an appropriate matrix to combine with the original to find a new matrix. The following operations could be performed to show these transformations.

Take the three vertices of a triangle. This can be written as a 2x3 matrix, called M. (These are of the two-dimensional variety. We can also transform three dimensional objects in a way similar to these examples.)

Rotation of 90 degrees counter-clockwise: $\begin{bmatrix} 0 & -1 \\ 1 & 0 \end{bmatrix} M$

Rotation of 270 degrees counter-clockwise: $\begin{bmatrix} 0 & 1 \\ -1 & 0 \end{bmatrix} M$

Reflection in the x-axis: $\begin{bmatrix} 1 & 0 \\ 0 & -1 \end{bmatrix} M$

A dilation of 3: 3M

Translation of 3 to the right and 2 down: $\begin{bmatrix} 3 & 3 & 3 \\ -2 & -2 & -2 \end{bmatrix} + M$

Counter-clockwise rotation of α degrees: $\begin{bmatrix} \cos(\alpha) & -\sin(\alpha) \\ \sin(\alpha) & \cos(\alpha) \end{bmatrix} M$

More than one operation can be done by composing two of the functions. The points could be rotated 90 degrees counter-clockwise and then reflected across the x-axis: $\begin{bmatrix} 1 & 0 \\ 0 & 1 \end{bmatrix} \left(\begin{bmatrix} 0 & 1 \\ 1 & 0 \end{bmatrix} M \right)$.

Inverses of matrices may also be used. A matrix M was multiplied by matrix T so it was rotated to end up at M'. However, it isn't known what that original matrix M was. This transforming matrix can be found by multiplying each side of the equation by the inverse of T.

$$T \cdot M = M' \rightarrow T^{-1} \cdot T \cdot M = T^{-1} \cdot M' \rightarrow M = T^{-1} \cdot M'$$

Video games and other applications dependent upon graphics create pictures whose parts are made up of polygons. These polygons are usually triangles and their vertices are ordered triples rather than ordered pairs. These polygons can then be moved by placing their vertices into a matrix and then using the appropriate transformation to find the new location.

Cryptography

Matrices can be used to encode messages so they are not shared with others. Various techniques have been employed throughout the ages to hide important communications. The use of matrices can do this quite well. The following method, developed by Lester Hill and known as the Hill Cipher, was developed in 1929.[34]

Every letter of the alphabet can be assigned a number. For simplicity sake, we'll let A = 1, B = 2, C = 3, ..., Z = 26. These numbers can now be written in matrix form. Let's use the simple message, "attack."

Now for an encoding matrix. Any square matrix that has an inverse can be used. Suppose $\begin{bmatrix} 2 & 5 \\ 1 & 3 \end{bmatrix}$ is chosen. Now multiplying these matrices will encode this message.

$$\begin{bmatrix} 2 & 5 \\ 1 & 3 \end{bmatrix}\begin{bmatrix} 1 & 20 & 20 \\ 1 & 3 & 11 \end{bmatrix} = \begin{bmatrix} 7 & 55 & 95 \\ 4 & 29 & 53 \end{bmatrix}$$

Then how is this message decoded by the receiving party? Having the encoding matrix, then finding the inverse, gives the receiver a decoding matrix. So, multiplying by the inverse, $\begin{bmatrix} 3 & -5 \\ -1 & 2 \end{bmatrix}$, gives the original message to the receiver.

All that is necessary to make this work is the original encoding matrix. The code could be changed by simply using a different encoding matrix. Since the goal is secrecy, there could be a different matrix used for each day of the week. This method has pretty much everything one would want in a coding system. It is not easily figured out by outsiders, easy to encode and decode by those on your side, and easily changed.

Vectors

Vectors show magnitude and direction. This can be expressed mathematically a number of ways. Vectors can be described trigonometrically, with ordered pairs, or geometrically with arrows.

For many real life situations, to know the magnitude is about all the information that is needed—weight, height, and temperature are scalar quantities in which direction isn't a factor. However, many times the direction is important as well. If a weather front is moving in, meteorologists would want to know not only the wind speed, but also the direction it is moving.

There are many examples in which one or more vectors act together on an object, affecting its motion.

1. An airplane flies into a wind that is at a 30° angle to the plane's motion. There is, of course, the force of gravity to deal with. Even if there is no wind, there is the force of friction from the air itself.

2. A boat is being propelled by its motor, but is going against a current.

3. Weather fronts can be represented by vectors. For an approaching front, it would be important to know how fast it is moving and the direction it is headed. Weather fronts can collide. When they do, bad things can happen. To predict the where, the when, and the strength of that collision depends on the vectors representing those fronts.

4. You are skiing down a hill. While you push with your ski poles, gravity is pulling straight down, there is a crosswind blowing, and there is friction between the snow surface and the skis.

5. Andrew is dragging Karen on a sled. His motion is influenced by the weight of Karen, friction, and the force being exerted by Andrew, who is pulling at an angle at 35° to horizontal.

To take this last example, suppose Andrew is pulling with a force of 3 newtons. Because it is at an angle, not all of that force is being used in the forward direction. The vector can be broken into its component parts. To get the amount of force in the horizontal direction means solving $\cos(35°) = \frac{x}{3}$.

This turns out to be 2.46 newtons of force.

Dot Product

The dot product of two vectors gives a single number; thus a scalar quantity. The formula for the dot product can be used to find the angle between two vectors. The dot product of vectors can be found with the following: $\mathbf{v} \cdot \mathbf{w} = |\mathbf{v}||\mathbf{w}|\cos(\theta)$. Also if $\mathbf{v}$ is written as the ordered pair (x_1, y_1) and $\mathbf{w}$ is (x_2, y_2), the dot product can be found with $\mathbf{v} \cdot \mathbf{w} = x_1 x_2 + y_1 y_2$.

Examples:

1. Staying with the above sledding situation, supposing Karen was dragged 100 meters, what total amount of work was done by Andrew?

Work done can be found by finding the dot product of force and distance vectors.

$\mathbf{W} = \mathbf{F} \cdot \mathbf{d} = |\mathbf{F}| \cdot |\mathbf{d}|\cos(\theta) = 3(100)(\cos(35°)) = 246$ newton-meters = 246 joules

2. Two boats left a dock and are now located as vectors represented by (1,1) and (3,4). What angle was between the boats when they left the dock?

$$1 \cdot 3 + 1 \cdot 4 = \sqrt{2} \cdot 5\cos(\theta) \quad \rightarrow \quad \frac{7}{5\sqrt{2}} = \cos(\theta) \quad \rightarrow \quad \theta = 8.13°$$

The dot product can be used to find other important information. Since $\cos(90°) = 0$, any two vectors whose dot product is zero must be perpendicular. In physics, power is the dot product of force and velocity vectors. As seen in a previous example, work is the dot product of force and distance vectors.

Cross Product

Cross product can be found with the formula $\mathbf{v}\mathbf{x}\mathbf{w} = |\mathbf{v}||\mathbf{w}|\sin(\theta)$.

Torque is an amount of force causing an object to rotate. Torque can be found by taking the cross product of the force acting on the object and the lever arm.

Example:

What is the magnitude of torque used to tighten a bolt if you generate 70 newtons of force on a 0.3 meter wrench for an angle of 60 degrees?

$|\mathbf{t}| = |\mathbf{r}| \times |\mathbf{F}| = 70(0.3)\sin(60°) = 18.2$ newton-meters

In addition to torque, there are other areas where the cross product can be important.

1. If the cross product gives a value of zero, it means that the vectors are the same or are parallel to each other.

2. Angular momentum is the cross product of linear momentum and distance.

3. The cross product of two vectors gives the area of the parallelogram spanned by those two vectors.

4. The volume of a parallelepiped (the three dimensional shape made up of six parallelograms) is found by using both the dot product and the cross product. If the parallelepiped is defined by the three-dimensional vectors $\mathbf{a}$, $\mathbf{b}$, and $\mathbf{c}$, the volume is found by using $V = |\mathbf{a} \cdot (\mathbf{b}\mathbf{x}\mathbf{c})|$.

5. A tetrahedron is a pyramid shape made by connecting any four non-coplanar points. If those points are represented by vectors **a**, **b**, **c**, and **d**, the volume can be found by using the dot product and the cross product:

$$\text{Volume} = \left| \frac{(a-d) \cdot (b-d) \times (c-d)}{6} \right|$$

IV

ADVANCED MATH

Solving Polynomial Equations

Advanced mathematics textbooks are filled with examples of polynomial equations. Second degree equations are fairly common. For example, the path of projectiles—footballs, rockets, bullets—can be written as second degree equations. Finding adequate real world examples of polynomials of higher degree, however, can be challenging.

Algebra texts will often have examples that lack realism. *The length of a swimming pool is 12 feet more than its width with a total area of 190 square feet. Set up a polynomial equation and solve to find the length and width of the pool.* One might think that if someone had the inside knowledge to know it was 12 feet longer than it was wide, he should be able to get the length and width without resorting to solving a polynomial equation. The problem has value as a word problem to stimulate problem solving skills, but it really isn't much of an application.

Math books will often state a problem like the following: *The gross national product from 2000 to 2015 can be found by the polynomial function $f(x) = 0.3x^4 + 2x^3 - 0.4x^2 + 2.3x + 1.8$. Use the function to find the gross national product in 2010.* The typical student is not going to see this as a viable application either. The student is going to think, "Where did this come from? Is it just made up by the people that wrote the math book?" Problems like this do little to help students see real world applications. There needs to be at least some explanation of where those functions come from. Although the process may be a bit of a mystery to students, functions can be found that model a set of data.

Models

With the right calculator or website, modeling equations can be found without too much difficulty. Finding a "curve of best fit" or a "regression curve" can be demonstrated to students so they can see where these equations come from. A person doing this would enter the ordered pairs along with what degree polynomial is sought. After inputting the information, the calculator output will determine a curve that best fits the data along with a value showing how close the curve matches the data.

For example, take the progressive world records set in the men's shot put. We will select roughly one each decade. In 1909, the world record was 15.54 meters, which can be written as the ordered pair (9, 15.54). Continuing in this manner gives the following set of ordered pairs: (9, 15.54), (28, 16.04), (32, 16.20), (48, 17.68), (53, 18.04), (60, 19.67), (73, 21.82), (85, 22.62), (90, 23.12).

The degree of the polynomial can be chosen. Choosing a linear regression for this data yields the equation: $f(x) = 0.1058x + 13.35$.

For a quadratic regression, the function is $g(x) = 0.0008179x^2 + 0.02178x + 14.97$.

For a cubic regression, the function is $h(x) = -0.00002602x^3 + 0.004686x^2 - 0.137x + 16.51$.

Rather than seeing functions magically appear in their textbook, students get a sense of how these come to be. These functions can now be used as a tool to examine the data and make predictions.

Any of these functions could then be used to find answers to questions such as, "What is a prediction for what the discus world record should have been in the year 2000?"

Let us look at what each of these functions say about that. The year 2000 would correspond to an x value of 100.

$f(100) = 23.93$
$g(100) = 25.327$
$h(100) = 23.65$

Here is a good example showing that predictions don't always come to pass. In reality, the record in the year 2000 was the same as it was in 1990. After being broken over fifty times in the 1900s, the latest record, at 23.12 meters, has lasted for over two decades now. It is one of the longest lasting world records in modern times.

From this example, students can learn that mathematics is a valuable tool form making predictions about the future. Additionally, they can learn that predictions don't always work out.

Roller Coasters

A polynomial equation can be used to model the path a roller coaster might take. The designer's plan for the coaster is drawn on a piece of graph paper. Suppose that four of the points planned for the roller coaster are (0,60), (20,120), (50,80), (65,30). A curve of best fit could be found for those points.

If choosing a third degree polynomial, a program can be used that finds: y = 0.00065x³ – 0.132x² + 5.38x + 60. This polynomial equation could be used in a variety of ways. Different values could be substituted for x to find the height at a number of other points. The equation could be graphed. If the student has knowledge of calculus, the first derivative could be found. This would then lead to finding relative maximums, minimums, and the slope of the roller coaster track at any desired point.

Perhaps not just any points are selected, but the locations of the peaks and the valleys of the proposed track are chosen. Suppose they are located at x = 25, 37, and 112. Those locations would have a slope of zero. The first derivative set equal to zero could look like this: f'(x) = (x – 25)(x – 37)(x – 112) = 0. Those binomials could be multiplied, and the anti-derivative found. This would result in a fourth degree equation representing the path of the roller coaster.

Finding Maxima and Minima

Differential calculus can be used to find the high and low points of a function. Most high school students would not be familiar with calculus, but the basic concept could still be presented. Doing so can show an application of solving polynomial equations. It could be explained that the derivative of any function can yield the rate of change at any point on a curve. Students can also be shown that places where those rates are equal to zero are likely to be locations of relative maximums or minimums.

The zeros of a polynomial function give important information about its graph. It is a way to find the highest point in the path of a projectile. In other situations, it could show when an object is traveling fastest, when an object is traveling slowest, the minimum cost to produce an item, a company's maximum profit, etc.

Road Load Power

Road Load Power is defined as the amount of power at the wheels that is needed to maintain a car's velocity. The Road Load Power is a major factor in determining the fuel economy of a car. Several factors come into play in finding its value. The totals of the forces acting on the car can be expressed as Force = a + bv + cv². Since Power = Force x Velocity, the formula is[1]:

Road Load Power = $(a + bv + cv^2)v = av + bv^2 + cv^3$

The variable v is the velocity of the car. The variables a, b, and c represent various constants for the car. These variables are determined by the amount of friction caused by the tires' contact with the ground, the working of the moving parts within the car itself, and air resistance.

Scientific Calculator Values

Scientific calculators find values of functions, often by adding terms of a series.[2] For example, for a calculator to find the cosine of 2 radians is to evaluate $1 - \frac{2}{2!} + \frac{2^4}{4!} - \frac{2^6}{6!} + \ldots$ On the other hand, to find the value of $\cos^{-1}(.8)$ would be, in essence, solving the polynomial equation $0.8 = 1 - \frac{x^2}{2!} + \frac{x^4}{4!} - \frac{x^6}{6!} + \frac{x^8}{8!} \rightarrow 0.8 = 0.0000248x^8 - .00139x^6 + .04166x^4 - 0.5x^2 + 1.$

Exponential Growth Functions and Logarithms

Calculations Before Calculators

Calculators have changed things for the better. Imagine computing $34.8 \div 1.9988$ or $\sqrt[4]{76}$ without one. It is now a fairly quick process. Before calculators, it was not. However, mathematicians had their ways of dealing with those tough arithmetic problems.

Square roots and cube roots were found by looking them up on tables—tables that others had laboriously generated. So how did they come up with those tables, anyway? Initially, there were computational methods involving repeated divisions. Later, series were used to find those values. An important step in making computations much easier came in the 1600s with John Napier's development of logarithms.

Roots could be found by using properties of logarithms. (Throughout this book, assume "log" to stand for a logarithm with a base of 10, and "ln" to be a logarithm with a base e.)

$$x = \sqrt[4]{76} \quad \rightarrow \quad \log(x) = 0.25\log(76) \quad \rightarrow \quad x = \text{antilog}(0.4702) = 2.9526$$

Multiplication and division problems can be simplified by use of logarithmic properties that can turn them into much easier addition and subtraction problems.

$$x = 34.8 \div 1.9987 \rightarrow \log(x) = \log(34.8) - \log(1.9987) \rightarrow x = 17.41131736$$

The values of logarithms were found by looking them up in tables. Amazingly, Napier and fellow mathematician Henry Briggs generated tables of logarithms up to a value of 20,000 with an accuracy to fourteen decimal places.[3]

Slide rules are devices that aid in computation by using a logarithmic scale. Until the hand held calculator came into general use in the 1970s, the slide rule had played the role of calculator for three hundred years.

Banking Formulas

Simple interest uses the formula I = prt. However, most institutions don't handle interest this way. They use some form of compound interest. Money in an account that is compounded annually means that after a year has elapsed, the new principal is the original amount plus the interest that was gained during the past year. This principal, P, can be compounded any number of times per year, n, at a particular interest rate, r (written as a decimal). We can find the amount of money after time t with the formula $A = P(1 + \frac{r}{n})^{nt}$.

Examples:

1. How much money would there be if $1,000 is invested at 2 percent, compounded semi-annually for 5 years?

$$A = 1000\left(1 + \frac{0.02}{2}\right)^{2(5)} = \$1,104.62$$

2. An account is compounded monthly with an interest rate of 5 percent. How much would be needed as a beginning balance in order to have $2,000 at the end of 3 years?

$$2{,}000 = P\left(1 + \frac{0.05}{12}\right)^{12(3)} \quad \rightarrow \quad 2{,}000 = P(1.161472) \quad \rightarrow \quad P = \$1{,}721.95$$

3. After 8 years, with your money being compounded quarterly, your account grew from $2,000 to $2,413. What was the interest rate?

$$2{,}413 = 2{,}000\left(1 + \frac{r}{4}\right)^{4(8)} \quad \rightarrow \quad \sqrt[32]{1.2065} = 1 + \frac{r}{4} \quad \rightarrow \quad r = 0.0235 \text{ or } 2.35\%$$

4. A bank account pays 3 percent interest, compounded annually. You want your $5,000 to grow to $7,000. How long will you have to wait?

$$7{,}000 = 5{,}000\left(1 + \frac{0.03}{1}\right)^{1t} \quad \rightarrow \quad \log(1.4) = t \cdot \log(1.03) \quad \rightarrow \quad t = 11.38 \text{ years}$$

There are many other practical equations used in the banking realm. The following paragraphs show a couple of them.[4]

5. If $100 each month is put under the mattress, after ten years there would be $12,000 available. However, if that money is gaining interest, this investment would do much better. A formula to compute the total amount (A), with a regular monthly deposit (d), at a monthly interest rate (r), for n months is:

$$A = d\left(\frac{(1 + r)^n - 1}{r}\right)$$

Instead of under a mattress, that money is invested at 6 percent annual interest (0.5 percent monthly) for 10 years (120 months), there would be $16,387.93.

6. Thinking of buying a house? A key item of concern would be what the amount of the monthly payments. Finding the monthly payment (P) on the balanced owed (B) can be found with a similar looking formula.

$$B = P\left(\frac{1-(1+r)^{-n}}{r}\right)$$

Radioactivity

Matter that spontaneously emits energy and subatomic particles is said to be radioactive. Elements that do so have a certain half-life—the amount of time it takes for half of the element to decay. A substance with a mass of 120 grams, with a half-life of 7 years, would have 60 grams after 7 years, 30 grams after 14 years, 15 grams after 21 years, and 7.5 grams after 28 years. If a substance has a half-life of h years, a formula to find the amount left, N, from a starting sample of N_o, after t years, can be written as: $N = N_o(0.5)^{\frac{t}{h}}$

Examples:
1. Radon-226 has a half-life of 1,600 years. There were 20 kilograms of it in the year 2000. How much of it will be left in the year 3535?

$$N = 20(0.5)^{\frac{1,535}{1,600}} = 10.29 \text{ kilograms}$$

2. How many years would it take the 20 kilograms of Radon-226 to reach a mass of 1 kilogram?

$$1 = 20(0.5)^{\frac{t}{1,600}} \quad \rightarrow \quad \log(0.05) = \frac{t}{1,600}\log(0.5) \quad \rightarrow \quad t = 6,915 \text{ years}$$

3. In 10 years, a substance decays from 23 kilograms to 21 kilograms. What is the half-life of the substance?

$$21 = 23(0.5)^{\frac{10}{h}} \quad \rightarrow \quad \log(0.9130) = \frac{10}{h}\log(0.5) \quad \rightarrow \quad h = 76.2 \text{ years}$$

In 1986, the nuclear accident at Chernobyl released 1,000 kilograms of Cesium-137 into the atmosphere. The half-life of Cesium-137 is 30.17 years.[5] There will be y kilograms present x years after the accident, as given by the equation: $y = 1,000(0.5)^{\frac{x}{30.17}}$.

4. How much will be present in 2026? [Answer: 398.92 kilograms]
5. How many years will it take until there are 50 kilograms remaining? [Answer: 130.4 years]

As seen in this example, half-lives can be found without waiting to see how long it takes for half of the substance to decay. Half-lives can range from minutes to billions of years. It gets a little trickier for those with longer half-lives, but essentially the same method is used.

Richter Scale

In the 1930s professor Charles Richter of Caltech developed a scale for measuring the magnitude of earthquakes. While a new scale has been developed in recent years, for most earthquakes the numbers end up being roughly the same. Richter's formula was $M = \log_{10}\left(\frac{I}{S}\right)$, where I is the intensity of the earthquake and, for comparison, S is a constant that is the intensity of a barely perceptible earthquake. Because a logarithm of base ten is being used, a Richter number one unit greater means an increase of ten in the actual intensity of the quake. An earthquake registering 6.0 is not twice as strong as a 3.0. Because $\frac{10^6}{10^3} = 10^3 = 1000$, a 6.0 quake is a thousand times as strong as a 3.0.

Examples:

1. An earthquake registers 4.2 on the Richter scale. What would be the Richter number associated with an earthquake that is 4 times as strong?

One way of looking at this problem is as follows:

Substituting 4.2 in for M, gives $4.2 = \log\left(\frac{I}{S}\right)$

Converting to exponential form, $\frac{I}{S} = 10^{4.2} \rightarrow \frac{I}{S} = 15{,}848.932$

An intensity four times as great gives $\frac{4I}{S} = 63{,}395.728$

Taking the log of each side gives the answer: $\log\left(\frac{4I}{S}\right) = 4.802$

2. How much stronger is an earthquake measuring 5.3 than a 4.1?

This problem comes down to finding the ratio of their intensities, I_1 and I_2.

Their equations can be written $5.3 = \log\left(\frac{I_1}{S}\right)$ and $4.1 = \log\left(\frac{I_2}{S}\right)$

These can be broken into $5.3 = \log(I_1) - \log(S)$ and $4.1 = \log(I_2) - \log(S)$

Subtracting these equations, we get $1.2 = \log(I_1) - \log(I_2)$ or $1.2 = \log\left(\frac{I_1}{I_2}\right)$

Changing to exponential form gives the answer $\frac{I_1}{I_2} = 10^{1.2} = 15.8$ times as strong.

To solve these in such a way follows the strict definition of the Richter scale and shows examples of logarithmic properties in action. However, some shortcuts can be taken to get the same answers. In the first problem, the Richter number was sought that corresponded to an earthquake 4 times as intense as a 4.2. It could be found by using the fact that increasing a Richter number by 1 is a ten-fold increase in the intensity of the earthquake. Thus, the following method could be used.

$$4(10^{4.2}) = 10^x \rightarrow \log(63{,}395.7) = \log(10^x) \rightarrow x = 4.802$$

For the second problem, the issue was to find out how much stronger a 5.3 earthquake is than a 4.1.

$$\frac{10^{5.3}}{10^{4.1}} = 10^{1.2} = 15.8 \text{ times as strong}$$

Loudness

The decibel is the common unit for measuring sound. It is one-tenth of a bel and was named in honor of Alexander Graham Bell. The formula for measuring sound is very similar to that of measuring earthquakes.

$D = 10 \cdot \log_{10}\left(\frac{P}{P_0}\right)$ D is the number of decibels

                          P is the intensity of the sound

                          P_0 is a constant and is the weakest sound humans can hear

Examples:

1. At take-off, a jet is estimated to have a loudness of 120 decibels. How much louder is this than a vacuum cleaner at 72 decibels?

It will be necessary to find the ratio of the intensity of these sounds, X and Y.

For the jet: $120 = 10 \cdot \log\left(\frac{X}{P_0}\right)$ $\rightarrow$ $10^{12} = \frac{X}{P_0}$ $\rightarrow$ $10^{12} \cdot P_0 = X$

For the vacuum: $72 = 10 \cdot \log\left(\frac{Y}{P_0}\right)$ $\rightarrow$ $10^{7.2} = \frac{Y}{P_0}$ $\rightarrow$ $10^{7.2} \cdot P_0 = Y$

The desired ratio: $\frac{X}{Y} = \frac{10^{12} \cdot P_0}{10^{7.2} \cdot P_0} = 10^{4.8} = 63{,}095.7$ times as intense a sound.

Again, it doesn't follow the exact definition, but students who are not familiar with logarithms can use the rules for exponents to obtain the same answer: $\frac{10^{12}}{10^{7.2}} = 10^{4.8} = 63{,}095.7$.

2. A lawn mower operates at approximately 90 decibels. What number of decibels would be twice as intense?

The loudness of the lawn mower can be represented with, $90 = 10 \cdot \log\left(\frac{P}{P_0}\right)$. The sound twice as loud could be written as $10 \cdot \log\left(\frac{2P}{P_0}\right) = 10 \cdot \log(2) + 10 \cdot \log\left(\frac{P}{P_0}\right) = 3.01 + 90 = 93$ decibels.

The value of the constant, P_0, is 10^{-12} watts/m². That value wasn't necessary in the above problems, but is needed in the following problem.

3. A concert has an intensity of 100 watts/m². How many decibels is this sound?

$$D = 10 \cdot \log\left(\frac{100}{10^{-12}}\right) = 10 \cdot \log(10^{14}) = 10 \cdot 14 = 140 \text{ decibels}$$

pH Numbers

A substance's pH number is a measure of its acidity (for acids) or alkalinity (for bases). Technically, a substance's pH number is based on its concentration of hydrogen ions. Since the concentration of hydrogen ions is awkward because it is so small, chemists use the concept of logarithm to define the pH number. It is defined by the equation pH $= - \log_{10}[H^+]$. (In chemistry talk, the concentration of hydrogen ions is written as $[H^+]$). The numbers range from 0 (very acidic) to 14 (very alkaline). Battery acid is a very strong acid. Its pH number is around 1. Water is neutral and has a pH of 7. Bleach is a very strong base with a pH around 12.5.

Examples:

1. If the pH number of vinegar is 3.5, what is the value of $[H^+]$?

$$3.5 = - \log[H^+] \to \text{antilog}(-3.5) = [H^+] \to [H^+] = 0.000316$$

Or this can also be found by converting to exponential form,

$$3.5 = - \log[H^+] \to 10^{-3.5} = 0.000316$$

2. A substance has a hydrogen ion concentration of 1.82×10^{-5}. What is its pH number?

Once again, there are a couple of ways to solve this problem.

$$pH = - \log(1.82 \times 10^{-5}) = - \log(0.0000182) = - (-4.74) = 4.74$$

Or using the rules of logarithms,

$$pH = - \log(1.82 \times 10^{-5}) = - (\log(1.82) + \log(10^{-5})) = - (0.26 + -5) = 4.74$$

Star Brightness

If the Sun and the star Alpha Centauri were the same distance from the Earth, they would each seem, roughly, equally bright to us. They don't, however, because the Sun is a lot closer. It could be said that they have the same absolute magnitude of brightness.

If all stars were the same distance away from us, we could rank their apparent brightness. Astronomers have chosen 10 parsecs to be that distance. (One parsec is 3.26 light years.)

The formula for comparing absolute magnitude (M) and apparent magnitude (m) and the distance (d) to the star is as follows[6]:

$$m - M = 5 \cdot \log\left(\frac{d}{10}\right)$$

Examples:

1. The Sun and Alpha Centauri have absolute magnitudes of 4.83 and 4.1, respectively. Their apparent magnitudes, though, are figured to be –26.8 and –0.3. (The brighter stars have lower apparent magnitudes.) How far is it from the Earth to Alpha Centauri?

$$-0.3 - 4.1 = 5 \cdot \log\left(\tfrac{x}{10}\right) \quad \rightarrow \quad -0.88 = \log\left(\tfrac{x}{10}\right) \quad \rightarrow \quad x = 1.32 \text{ parsecs}$$

2. The star Rigel has an apparent magnitude of 0.14 and is 276.1 parsecs away from Earth. What is the absolute magnitude of Rigel? [Answer: –7.1]

Natural Logarithms and e

Bell-Shaped Curves

Many sets of statistics are normally distributed. For any randomly chosen group of people, their heights, weights, IQs, how fast they can run a mile, etc., are all likely to have a normal distribution. If the data were graphed, the result would be a bell-shaped curve. If we were to look at the heights of all adult females, this bell shape would most likely be seen. A characteristic of this shape would be that the farther away from the mean average, the fewer women there would be. There are far more women in the range from 5'6" to 5'8" than there are from 6'6" to 6'8". The equation that describes this bell-shaped curve is:

$Y = \frac{1}{\sigma\sqrt{2\pi}} e^{-\frac{1}{2}\left(\frac{x-\mu}{\sigma}\right)^2}$ where σ is the standard deviation and μ is the mean average.

Examples:

1. IQ scores are very nearly a normal distribution. For the general population, an IQ score of 100 is considered to be average, with a standard deviation of approximately 15.[7] How would an equation be written to describe this distribution?

$$Y = \frac{1}{15\sqrt{2\pi}} e^{-\frac{1}{2}\left(\frac{x-100}{15}\right)^2}$$

2. Similarly, an equation could be written for this fact: In 2012, students taking the mathematics portion of the SAT had a mean average of 496 with a standard deviation of 114.[8]

This is the curve that is the basis of the phrase "grading on the curve." Many have an understanding of grading on the curve that isn't entirely accurate. For some, it means the highest score is taken to be 100 percent and other scores are moved up accordingly. That is a fine thing to do, but isn't really grading on the curve. Students also assume that grading on the curve will give everyone higher grades. Again, not so. To grade on the curve means grades will be determined by where they fall when the scores are normally distributed. The A grades might only be those that are more than one standard deviation above the mean (the top 16 percent of the class). If this is the case, assume that scores on a recent test had a mean average of 63 and a standard deviation of 11. If A's go to those students that are greater than one standard deviation above the mean, then those with scores above 74 will receive an A.

Continuously Compounded Interest

We have already visited the concept of compound interest. A bank could offer an interest rate that is compounded annually, monthly, daily, or, if desired, even by the second. Letting the number of times the interest is compounded approach infinity is the concept of continuously compounded interest. While banks don't tend to offer this, they could. The formula is actually more compact than the one used for other forms of compounding. The formula is $A = Pe^{rt}$, where P is the principal, r is the interest rate (as a decimal), and t is the time in years.

One practical use for this concept is that it quickly gives an upper bound for what various other compoundings would yield. This is especially helpful since the number of times the principal is compounded doesn't tend to make a big difference on how much interest is gained anyway. A good exercise would be for students to take a certain situation and find how much would be made with different compoundings. They would quickly learn that the principal and the interest rate have a lot more to do with how much money is made than the number of times it is compounded.

For example, $1,000 compounded annually at 3 percent interest after 5 years would give $y = 1,000(1 + .03)^5 = \$1,159.27$. The same principal, compounded continuously, would give $y = 1,000e^{5 \cdot 0.03} = \$1,161.83$. That is a modest gain of $2.56.

Examples:

1. How long would it take for $1,500, if compounded continuously at 3.5 percent, to grow to $2,400?

$$2,400 = 1,500e^{0.035x} \rightarrow \ln(1.6) = 0.035x \rightarrow x = 13.43 \text{ years}$$

2. In 4 years, $2,000 grew to $2,850. Assuming continuously compounded interest, what was the interest rate during this time? [Answer: 8.9 percent interest]

The Rule of 72

If an amount of money is invested, how long would it take for it to double? Equations can be used to find out. However, the Rule of 72 can be used to give a good, quick estimate. To use the Rule of 72, divide 72 by the interest rate to get the approximate time it would take the money to double. By using the Rule of 72, money invested at 6 percent would take about 12 ($72 \div 6$) years to double. Money invested at 10 percent would take about 7.2 ($72 \div 10$) years to double.

Using the formula for compound interest, and assuming the interest is compounded annually, gives the formula $A = P(1 + r)^t$. At 5 percent interest, compounded continuously, how long would it take the principal to double?

$$2P = Pe^{0.05t} \rightarrow 2 = e^{0.05t} \rightarrow \ln(2) = 0.05t \rightarrow t = 13.86 \text{ years}$$

The rule of 72 would say that at 5 percent interest it would take $(72 \div 5) = 14.4$ years. A pretty good estimate.

The Rule of 72 gives other good estimates. At 10 percent interest, it takes 6.9 years to double. The Rule of 72 says 7.2 years. At 12 percent it would take 6.1 years. The Rule of 72 gives a prediction of 6.0 years.

Some people use a Rule of 70 or a Rule of 69. These will also give good approximations. The number 72 has the advantage of being divisible by several numbers. Also, one or the other might be a bit more accurate depending on how the interest is compounded. Why do these numbers give good approximations, anyway?

We want to see what interest rate causes the principal to double, i.e., to go from a value of P to 2P. Taking the continuously compounded interest case:

$$2P = Pe^{rt} \quad \rightarrow \quad \ln(2) = \ln(e)^{rt} \quad \rightarrow \quad \ln(2) = rt \quad \rightarrow \quad \frac{0.693}{r} = 2$$

So, for the compounded continuously case, dividing 69.3 by the interest rate percentage would give the number of years it takes to double.

Newton's Law of Cooling

Your mother might tell you to come get your dinner before it gets cold. More accurately, your mother might tell you to come get your dinner before it reaches the ambient room temperature. Once food is removed from a hot stove, it will begin to cool off. But how quickly? Isaac Newton answered that question with the formula now known as Newton's Law of Cooling.

$T = T_s + (T_0 - T_s)e^{-kt}$, where T_0 is the initial temperature of the object
T_s is the surrounding temperature
k is a constant
T is the temperature after time t has elapsed

Examples:

1. A bowl of soup is 120° F. Room temperature is 72° F. If k has a value of 0.37, what will the soup's temperature be in half an hour?

$$T = 72 + (120-72)e^{-0.37(0.5)} = 111.9°$$

2. A metal is heated to a temperature of 250° F. If the room temperature is 75° and k = 0.16, how many minutes will it take the metal to reach a touchable 100°?

$100 = 75 + (250–75)e^{-0.16t} \rightarrow \ln(0.143) = \ln(e^{-0.16t}) \rightarrow t = 12.2$ min

3. From the files of *CSI*, there is a murder scene in which a body is discovered in a 68° motel room. The body is found to have a temperature of 80°. An hour later, the body has a temperature of 75°. How long ago was the time of death?

First, we can substitute values in order to find k.

$75 = 68 + (80–68)e^{-k(1)} \rightarrow \ln(0.5833) = \ln(e^{-k(1)}) \rightarrow k = 0.539$

We will assume that, when last alive, the body had a temperature of 98.6°. Numbers can now be substituted to find the time of death.

$80 = 68 + (98.6–68)e^{-0.539(t)} \rightarrow t = 1.74$ hours

It looks like the individual died almost approximately an hour forty-five minutes before the recorded 80 degree temperature.

Population Growth and Decay

A formula used to model population is $A = Ce^{rt}$, where C is an initial population that grows to the amount A after time t. The letter r is the rate of growth.

Examples:

1. In 7 years a city grows from 20,000 to 23,500 people. What is the value of r?

$23,500 = 20,000e^{7r} \rightarrow \ln(1.175) = 7r \rightarrow r = 0.02304$

2. What would be a prediction for the city's population in another two years?

$A = 20,000e^{(0.02304)9} = 24,608$

3. Using this model, how many years would it take for the city's population to reach 30,000?

$30,000 = 20,000e^{(0.02304)t} \rightarrow \ln(1.5) = 0.02304t \rightarrow t = 17.6$ years

4. According to the U.S. Census, the population of the United States in 1910 was 92.2 million people. The population in 1930 was 123.2 million. Based on these numbers, what would be an estimate for the population of the United States in 1960?

The equation $123.2 = 92.2e^{20r}$ could first be solved to find the value for r. Once found, that value could then be substituted to find the population in 1960 (t = 50). The answer turns out to be 190.3 million, which is quite close to 1960s actual population of 189.3 million. Granted, some good luck is involved in coming this close to the actual value. Because of societal factors such as depressions, wars, baby booms, etc., rates of growth could vary quite a bit over the years.

Students can be confused by growth and decay formulas that sometimes make use of the value "e" and sometimes do not. While the equations look quite a bit different, they give the same results. To make a comparison, let's take one application and see how it can be dealt with both ways.

Tritium is a radioactive form of hydrogen, having a proton and two neutrons in its nucleus. Because it can cause phosphors to glow, it is used in exit signs, on gun sights, and on watch dials. It has a half-life of 12.3 years. Because of this decay, one might notice objects such as watch dials harder to see as time goes by.

Suppose there are 100 ounces of tritium. How much will be left after 5 years? This can be solved in two different ways. We'll use more decimal places than we normally might, just to be sure these methods both give the same answer.

We have already seen that one formula which can be used is
$N = N_o(0.5)^{\frac{t}{h}}$

$$N = 100(0.5)^{\frac{5}{12.3}} = 75.44493402 \text{ ounces.}$$

A second way is to use the formula $A = Ce^{rt}$. First, tritium's half-life can be used to find the value of r:

$$A = Ce^{rt} \rightarrow 50 = 100e^{12.3r} \rightarrow r = -0.056353429.$$

Now to deal with our specific situation,

$$A = Ce^{rt} \rightarrow A = 100e^{-0.056353429 \cdot 5} \rightarrow A = 75.44493402 \text{ ounces.}$$

Nice.

Capacitance

An R-C circuit is an electrical circuit composed of resistors and capacitors and is driven by a voltage source such as a battery. The capacitance is a measure of how well the capacitors store charges. It can be found by looking at its initial charge (V_0) when there is a battery source, then examining the voltage (V) at a time (t) when that source is turned off. By substituting the value of the resistor (R), the capacitance (C) can be found with the formula[9]:

$$V = V_0\left(e^{\frac{-t}{RC}}\right)$$

Logistic Growth

Many populations exhibit a growth rate that follows an exponential model. However, there are situations in which populations appear to be growing exponentially, but have constraints that eventually limit that growth. These situations are often described by a logistic growth model.

Deer are introduced into an area. Their population grows, but as it grows, they overgraze the area to the point that there is not enough food to sustain the herd, putting a limit on the growth.

A flu epidemic is sweeping through a school. More and more people are getting the flu as people come into contact with others that are sick. At some point the rate of infection will slow down. In part, this is because, as time goes by, it is more and more likely the infected people's contacts are with people that are already infected. Another limiting factor is the number of students in the school, because there obviously can't be more cases than students.

The spreading of rumors can follow this model. In some people's mind, they will feel the need to spread the information they have. They tell some people who then tell others. The rumor may spread rapidly, but this rapid growth will begin to slow down. Those spreading the gossip will eventually tell the rumors to people who already know, or to those who won't participate. Thus, there is a limit to this growth.

The equation of a logistic curve can be written in the form

$$y = \frac{c}{1 + ae^{-rx}}.$$

Examples:

Suppose the equation $y = \frac{98}{1 + 193e^{-0.3t}}$ roughly describes the square inches area covered by bacteria growing on meatloaf left in the back of your refrigerator for t days.

1. How many square centimeters would be covered in 5 days? [Answer: 2.2 square inches]

2. After how many days will there be 50 square inches of coverage? [Answer: 17.7 days]

3. As t gets larger and larger, what is the largest possibility for the value of y? [Answer: 98 sq. inches]

Carbon-14 Dating

All living things contain carbon-12 and carbon-14. We living things primarily contain carbon-12. It is called carbon-12 because it has 6 protons and 6 neutrons in its nucleus. Because of the action of cosmic rays, life forms can pick up carbon-14 which contains 6 protons and 8 neutrons. Carbon-14 is radioactive, but carbon-12 is not.[10]

All living things have the same ratio of carbon-12 to carbon-14. They keep that ratio as long as they live. However, when they die, the carbon-12 stays and the carbon-14 dissipates as part of its radioactive decay process.

Carbon-14 has a half-life of 5,730 years. So, if a sample had half as much carbon-14 it would be expected to have for the amount of carbon-12, it must have passed on 5,730 years ago.

Other time periods can be determined by the following formula:

$$t = \left(\frac{\ln(r)}{-0.693}\right) \cdot 5{,}730$$

The value r is the ratio of the current amount of carbon-14 compared to the amount of carbon-14 that would have been found when in living tissue.

Examples:

1. A sample has 23 percent of the carbon-14 that it would be expected to have. When did it die?

$$t = \left(\frac{\ln(0.23)}{-0.693}\right) \cdot 5{,}730 = 12{,}152 \text{ years ago}$$

2. A bone is found from an animal that had died 250 years ago. What percent of its original amount of carbon-14 should be present? [Answer: 97.02 percent]

3. A human bone is found at a construction site. After testing at a lab, it is determined that the carbon-14 ratio is 95.3 percent of what it would be when living. How long ago did this person pass away? [Answer: 398 years ago]

The point was made previously that it can be confusing that some exponential growth problems use the value e and some do not. Some basic knowledge of logarithmic properties can show that, in this case of carbon-14 dating, those different forms are again equivalent.

So, how is the formula $N = N_0(0.5)^{\frac{t}{h}}$ equivalent to the $t = \left(\frac{\ln(r)}{-0.693}\right) \cdot$ 5,730 formula used for carbon-14 dating?

Starting with $N = N_0(0.5)^{\frac{t}{h}}$, and manipulating the equation:

$$N = N_0(0.5)^{\frac{t}{h}} \quad \rightarrow \quad \frac{N}{N_0} = (0.5)^{\frac{t}{h}} \quad \rightarrow \quad \ln\left(\frac{N}{N_0}\right) = \ln(0.5)^{\frac{t}{h}}$$

Since r was defined as the ratio of the carbons, and h for carbon is 5,730, substitutions can be made.

$$\ln(r) = \ln(0.5)^{\frac{t}{5{,}730}} \quad \rightarrow \quad \ln(r) = \frac{t}{5{,}730}\ln(0.5) \quad \rightarrow \quad \ln(r) = \frac{t}{5{,}730}(-0.693)$$

Now, doing a little algebra to get t by itself, gives the other formula: t = $\left(\frac{\ln(r)}{-0.693}\right) \cdot 5{,}730$

Hyperbolic Functions

There are six hyperbolic functions, corresponding to the six trigonometric functions. Each of their definitions makes use of the number e. For example, $\cosh(x) = \frac{e^x + e^{-x}}{2}$ and $\tanh(x) = \frac{e^x - e^{-x}}{e^x + e^{-x}}$. A hanging rope or chain can be described with a hyperbolic function. This shape is called a catenary and has an equation in the form $y = a \cdot \cosh\left(\frac{x}{a}\right) + b$. The catenary shape can be seen in the shape of the power lines along a road. The cable on a suspension bridge with the weight of the attached road forms a parabola, although without the weight of the road, the shape would be a catenary.

An important feature of the catenary shape is that tension along its path is minimized. Because of this, they have a number of engineering applications.

If this shape is turned upside down, it looks like an arch. In fact, the catenary is the strongest possible type of arch. These catenary arches have long provided supporting strength for hallways and giant cathedrals because this shape evenly distributes the forces pulling down along the arch. The Gateway Arch in St. Louis is a type of catenary. It can be modeled by the equation[11] $y = 693.8597 - 68.7672 \cdot \cosh(0.0100333x)$.

If a catenary is rotated, it forms a dome. The catenary shape was employed in the first domed sports stadium, the Houston Astrodome, built in 1965. Many other domed buildings since then have employed this shape because of its strength.

Rocket Man

Called the Ideal Rocket Equation,[12] $\Delta v = v \cdot \ln\left(\frac{m_0}{m_1}\right)$, gives the maximum amount of velocity that can be attained by a rocket. An astronaut is going to blast off in a rocket with mass m_0. Later, having used some of its fuel, the rocket will have a mass of m_1. If the fuel is expelled with a velocity of v, our rocket man can expect to increase his velocity by Δv.

Complex Numbers

Complex numbers are of the form a + bi, with a and b being real numbers and i standing for $\sqrt{-1}$.

Historically, there has been resistance to "new" types of numbers. New concepts such as negative numbers and irrationals were not universally accepted. This was also true with the proponents of imaginary numbers. It is not just happenstance that some decided they should be referred to as "imaginary." Students are often chagrined by the fact that you cannot hold onto 3i objects or have a team with 7i players. This is true, but you also can't have –12.7 people or eat $\sqrt{7}$ cookies. It is possible to find situations in which negatives, radicals, and imaginaries have meaning.

Mathematics

Perhaps not strictly a real world application, but complex numbers are important in giving a completeness to solving equations. Without negative numbers, there is no solution to the equation x + 7 = 4. Without radicals there is no solution to $x^4 = 123$. Without imaginary numbers there is no solution to $x^2 = -5$.

Without complex numbers, there is no Fundamental Theorem of Algebra. This theorem states that, given an nth degree polynomial equation, there must be n solutions. If limited to the real numbers, we can't be sure how many solutions an equation might have. However, if by including the imaginary numbers, any fifth degree equation has five solutions. Finding them might be a bit of a challenge, but they are there. It is, mathematically speaking, very satisfying.

Electricity

Complex numbers are rarely used when dealing with direct current, but commonly used with alternating current. In alternating current, as opposed to direct, the flow of electric charge periodically reverses direction. Because of this, many aspects of alternating current cannot be described with only one number. Real number ordered pairs could be used, but due to the sinusoidal nature of alternating current, there are advantages to using complex numbers. For any number plotted on the complex plane, multiplying by $\sqrt{-1}$ gives a new complex number rotated counterclockwise 90° from the original. Multiplying by $-\sqrt{-1}$ is a rotation of 90° in the clockwise direction from the original. This turns out to be very helpful and is one reason electrical concepts often have values expressed as complex numbers. Following are some of the ways they make their appearance.

An inductor is a device in an electrical circuit that stores energy in a magnetic field. Circuits can contain both resistance and inductive reactance. Both are measured in ohms and resist the flow of electricity. Electrical engineers express this as a complex number. The resistance, R, is the real number portion, and the inductive reactance, X_L, the imaginary portion. It is written as $R + jX_L$, where $j = \sqrt{-1}$. (Since, in the world of electricity, the letter i represents current, electrical engineers use the letter j as the imaginary unit.)

Like inductors, a capacitor is a device that stores energy, but in this case, it does so in an electric field. The capacitive reactance, X_C, and resistance, R, can both be present in a circuit. This is expressed as a complex number, $R + jX_C$. With either inductors or capacitors, these values can be plotted on a complex plane with the resistance being on the real number x-axis, and the capacitive reactance on the imaginary y-axis.

Voltage and current can also be written as complex numbers. For example, voltage across a resistor is taken to be a real quantity, and the voltage across an inductor is taken to be an imaginary quantity.

Examples:

1. A circuit has a voltage represented 2 – j4. What is the voltage after a counterclockwise rotation of 90 degrees?

$$(2 - j4) \cdot j = 2j - j^24 = 4 + j2$$

2. A circuit has a voltage represented by –3 + j2. A counterclockwise rotation of 45 degrees is accomplished by multiplying by 1 + j.

$$(-3 + j2)(1 + j) = -3 - j3 + j2 + j^22 = -5 - j$$

Reactance is similar to resistance except that reactance varies depending on the frequency while resistance does not. The impedance in a circuit is the sum of the reactance and the resistance. All are measured in ohms.

3. The impedance in one part of an electrical circuit is 6 + j2. In another part it is 3 – j5. What is the total impedance? This calls for complex number addition.

$$(6 + j2) + (3 - j5) = 9 - j3$$

4. What is the voltage of a circuit with impedance of 4 + j5 ohms and current of 3 + j4 amps?

Ohm's Law states that the voltage is equal to the product of the current and the impedance. So,

$$V = (4 + j5)(3 + j4) = 12 + j31 + j^220 = -8 + j31 \text{ volts}$$

5. A circuit has a voltage of 12 + j3 and impedance of 2 – j3 ohms. What is its current?

$$12 + j3 = x(2 - j3) \rightarrow \frac{12 + j3}{2 - j3} = x \rightarrow \frac{2 + j3}{2 + j3} \cdot \frac{12 + j3}{2 - j3} = x \rightarrow x = \frac{15}{13} + j\frac{42}{13}$$

Quantum Theory

In his book, *A Brief History of Time*, Stephen Hawking spoke of how imaginary numbers are used in relativity theory. Physicist Richard Feynman developed what he called a sum-over-paths approach to describe quantum theory. In this approach, rather than following a single path or history, a subatomic particle follows a sum of every possible path it might take in space-time. As Hawking says, "To avoid the technical difficulties with Feynman's sum over histories, one must use imaginary time. That is to say, for the purposes of the calculation, one must measure time using imaginary numbers, rather than real ones. This has an interesting effect on space-time: the distinction between time and space disappears completely."[13]

While on the topic of quantum theory, when electrons make a "quantum leap," energy is produced. This affects all parts of the electromagnetic spectrum. Radio and television waves, x-rays, and visible light are all parts of this spectrum. Values are sometimes written as complex numbers with the electrical part being the real component and the magnetic part, the imaginary component.

Fractals

Fractals display fascinating shapes. Zooming in on a fractal shows the original shape being repeated ad infinitum. They can be created by substituting values into a function and obtaining a result, which is then substituted back into the function, giving a second result. That process continues. These continued substitutions, or iterations, can be graphed on the complex plane. Those values that produce iterations that remain close to the origin are part of the set. Plotting those points that converge produces the graph. As an added bonus, the rate at which those points converge could be set to correspond to various colors, making those cool posters.

This repeating fractal pattern doesn't take place when only real numbers are used. So, a knowledge of complex numbers, and how to perform calculations with them, is necessary.

The process can be seen in the following example.

Start with the function $f(x) = 2x + (1-3i)$ and a first value of $x = 0$.

$$f(0) = 2(0) + (1-3i) = 1-3i$$
$$f(1-3i) = 2(1-3i) + (1-3i) = 3-9i$$
$$f(3-9i) = 2(3-9i) + (1-3i) = 7-21i$$

… and so on. In this case, these iterations are getting farther and farther from the origin. So, $1-3i$ is not part of the set.

If the iterations did converge toward the origin, then the complex number $1-3i$ would be part of the set and would be plotted as a point on the graph.

There is a specific fractal called a Mandelbrot set, after mathematician Benoit (pronounced "Benwah") Mandelbrot (1924–2010). The Mandelbrot set is generated by the function $f(x) = x^2 + c$. The task is to find the values of c whose distances to the origin converge. To test whether a specific number c is part of the set, begin by finding $f(0)$ and see whether the subsequent iterations converge. For an example, the value $c = 2 + i$ could be tested. First, $f(0)$ is found and then used as the next value of x. The process is continued to find the iterations of the function $f(x) = x^2 + 2 + i$.

$$f(0) = 0^2 + 2 + i = 2 + i$$
$$f(2 + i) = (2 + i)^2 + 2 + i = 4 + 4i + i^2 + 2 + i = 5 + 5i$$
$$f(5 + 5i) = (5 + 5i)^2 + 2 + i = 25 + 50i + 25i^2 + 2 + i = 2 + 51i$$

These distances do not seem to be converging. The distance from the origin to $2 + i$ is $\sqrt{2^2 + 1^2}$ = 2.23. From the origin to $5 + 5i$ is $\sqrt{5^2 + 5^2}$ = 7.07. From the origin to $2 + 51i$ is $\sqrt{2^2 + 51^2}$ = 51.04. In fact, once the distance from the origin is greater than 2, the iterations are not going to converge. So $2 + i$ does not get to be part of the Mandelbrot set. Computers can do this work far faster than humans, but to try a few of these is certainly good practice for students.

As mentioned, the fractals exhibit a repeating pattern that continues to be seen as we zoom in on various parts. This repeating pattern is often seen in art and in nature. Arteries branch off, eventually becoming small capillaries. Other branching structures include the limbs of a tree, the tracheal tubes in the lungs, river systems, and snowflakes. We also see this effect by zooming in on coastlines. Fractals can be used to estimate the length of a coastline or a river. Fractals have been used to create landscapes of faraway planets for science fiction movies.[14] To save space on computer hard drives, fractal compression is used to compact data.[15] Through repeated iterations, the distance between data points are shrunk to a minimum.

Polar Coordinates

A polar coordinate system is a way to graph ordered pairs; just like its cousin, the rectangular coordinate system. While most high school mathematics is done in the rectangular system, many equations and their graphs are much easier to deal with using polar coordinates. For example, spirals, cardioids, limaçons, and flower shapes are much easier handled in the polar system. A flower shape, known as a rose curve, can be written in the form r = acos(nθ) in the polar coordinate system. It is quite a bit messier in the rectangular system.

Microphones

Microphones can be constructed to pick up sounds in different directions.[16] An omni-directional microphone picks up sounds a certain distance from the mike. To graph its range would show a circle with the microphone at the center. Circles aren't too much of a problem in either system. However, the graph showing the range of other types of microphones is more easily accomplished with polar coordinates.

A cardioid microphone picks up sounds in front of, and on the sides, but not behind a microphone. Its mathematical shape is called a cardioid. It might be used to pick up sounds of a choir while minimizing sounds coming from the audience. A shotgun microphone is similar to the cardioid mike, but is much more aimed toward one direction. A bi-directional microphone primarily picks up sounds in two opposite directions; being excellent for two-person interview situations. The graphs for all of these are best handled in the polar coordinate system.

The Golden Spiral

Spirals can be found in a various areas, from the pattern of the seeds in a tiny flower to that of galaxies. These shapes can combine what would seem to be generally unconnected mathematical concepts—spirals, logarithms, the value e, and the golden ratio. Many spirals in nature are what are known as logarithmic spirals. They are of the form $r = ae^{b\theta}$, where (r,θ) are the collection of ordered pairs and a and b are constants. A specific logarithmic spiral is the golden spiral. Every 90°, the distance from the center increases by a 1.618—which happens to be the golden ratio.[17] How cool is that? The equation for the golden spiral is $r = ae^{0.30635\theta}$.

Radar

Radar finds objects by sending out signals and seeing if they return. If they return, it must have bounced off something. Radar waves permeate air, but not more solid objects. Sonar functions similarly, only underwater. The concepts of radar and sonar allow a submarine to locate another boat, air traffic controllers to locate planes, and a policeman to check out the speed of a car.

Radar marks locations by the angle and distance from its antennae. This, of course, is essentially how the polar coordinate system works. A radar screen is basically a piece of polar coordinate graph paper.

Polar Regions

A map of the United States showing latitude and longitude will appear to be based on the rectangular coordinate system. However, viewing a map of the North or South Pole looks more like the polar coordinate system. A useful exercise might be to use a map of Antarctica and list various points in terms of their latitude and longitude. Then those same locations could be written in polar coordinate form. This could be done by using the scale provided on the map and listing the locations by (r,θ). The value of r could be found by using the map scale and, with a ruler, finding the distance from the South Pole. The value of θ could be the number of degrees from the Prime Meridian.

Electronics

As previously mentioned, values associated with alternating current are often written as complex numbers. Electrical engineers find that those complex numbers can, at times, be better expressed either in rectangular or polar coordinate form, depending on the need.

Example:

The current is found to be 3 + j4. This can be transformed to polar coordinate form by using the Pythagorean Theorem and some basic trigonometry.

$$\sqrt{3^2 + 4^2} = 5$$

$$\tan(A) = \frac{4}{3} = 53.1°$$

In polar coordinates, this is the point (5, 53.1°)

Sequences and Series

A sequence is any ordered list of numbers, with or without a pattern. However, we are typically most interested in those that do have a pattern.

Arithmetic Sequences and Series

An arithmetic sequence adds a common difference, d, to a term to find the next term in the sequence.

The nth term of a arithmetic sequence is found with the formula $a_n = a + (n - 1)d$

The sum of the first n terms of an arithmetic series is found with $S_n = \frac{n(a + a_n)}{2}$

Examples:

1. A pyramid in Egypt has, on one side, a bottom row containing 210 stones. On each higher level, there is one less stone. How many stones are there on that side of the pyramid?

$$S_{210} = \frac{210(210 + 1)}{2} = 22,155 \text{ stones}$$

2. If a stone is dropped from the top of a cliff, it will cover more ground each successive second. These distances will increase arithmetically. The first second it drops 16 feet, the next second it drops 48 feet, then falls 80 feet the third second. For each successive second, the stone increases its amount fallen by 32 feet. How many feet would it drop during the seventh second? How far would the stone fall for the full seven seconds?

$$A_7 = a + (n - 1)d = 16 + (7-1)32 = 208 \text{ feet}$$

The 208 feet represents just what was covered during the seventh second. The grand total of the seven second descent is the sum of arithmetic series, 16 + 48 + ... + 208.

$$S_7 = \frac{n(a + a_7)}{2} = \frac{7(16 + 208)}{2} = 784 \text{ feet}$$

Incidentally, another way to find this solution is to use the formula for falling objects, $f(t) = 16t^2$. Using this function, $f(7) = 784$ feet.

3. The sum of the interior angles of a triangle is 180°. The sum of the interior angles of a quadrilateral is 360°. The sum of the angles increases 180° every time add an additional side is added to the polygon. Thus, this describes an arithmetic sequence. What is the sum of the interior angles of a 16-sided figure?

Because the sequence started with a three-sided figure, a sixteen-sided figure is actually the fourteenth term of the sequence. So, $A_{14} = 180 + (14-1)180 = 2,520$ degrees.

Geometric Sequences and Series

A geometric sequence is one in which each term is multiplied by a fixed amount, r, to obtain the next value in the sequence. Sequences can be thought of as the discrete case of many of the mathematical situations we have seen already. Elsewhere in this book we have used the formula for compound interest. That formula can be used to solve problems such as finding the amount of money that can be gained by starting with $100 and compounding it annually at 10 percent interest.

This situation can also be thought of as a geometric sequence. The sequence 100, 110, 121, 133.1, 146.41,... represents a geometric series with r = 1.1. Thus, situations arising from exponential equations—money growing in an investment, radioactive decay, the spread of a disease, or the growth of a state's population can also make use of the concept of geometric sequences.

The nth term of a geometric sequence can be found with $a_n = ar^{n-1}$.
The sum of n terms of geometric series is found with $S_n = (a-ar^n)/(1-r)$.
The sum of an infinite number of terms can be found with $S = a/(1-r)$, if $|r| < 1$

Examples:

1. Radioactivity can be measured by half-lives. For example, radium–226 has a half-life of 1,620 years. It could also be viewed as an annual percentage decrease. How much of five pounds is left after three half-lives?

This problem can be solved by finding the nth term in a sequence, where n is the number of half-lives. Three half-lives from now will be the fourth term of the sequence and can be found with:

$$a_4 = 5(0.5)^{4-1} = 0.625 \text{ pounds}$$

There is another way of looking at this problem. Rather than n standing for the number of half-lives, n could stand for the number of years elapsed.

So, first it must be determined what the value of r would be. That can be found by observing that it takes 1,620 years for 1 pound to become 0.5 pounds. What is the yearly rate of decay for a substance with a half-life is 1,620 years? That value can be found by solving the equation:

$$0.5 = 1(x)^{1,621-1} \rightarrow -0.693147 = 1,620\ln(x) \rightarrow x = 0.9995722$$

From one year to the next, 99.95722 percent of the substance is left.

Now this problem can be solved. The three half-lives of Radium 226 is equivalent to $(3 \cdot 1,620 =)$ 4,860 years. This corresponds to finding the 4,861st term in the geometric sequence starting at 5 and having a common ratio of 0.9995722.

$$a_{4,861} = 5(0.9995722)^{4,861-1} = 0.625 \text{ pounds}$$

2. From 2011 to 2013 the United States grew at 0.7 percent each year. If that pace continues and the 2013 population of the U.S. is 315.1 million, what is the predicted population in the year 2050?

$$a_{38} = 315.1(1.007)^{38-1} = 407.89 \text{ million}$$

3. A basketball is dropped from a height of 20 feet. If it rebounds to 90 percent of its previous height each bounce, what height does it reach on its sixth bounce?

The ball starts at 20 feet and then has 6 bounces, so what is needed is the seventh term of the sequence.

$$a_7 = 20(0.9)^{7-1} = 10.629 \text{ feet}$$

4. Continuing with the previous situation; how far does this ball travel at the point that it hit the floor the seventh time?

$$S_7 = \frac{20 - 20 \cdot 0.9^7}{1 - 0.9} = 104.341$$

But not so fast. The ball (except right at the start) got to go up and down on each bounce. So each bounce height will have to be doubled except for the starting amount. Its total distance travelled is 2(104.341) – 20 = 188.682 feet.

5. The laws of physics will cause the ball in the previous example to eventually come to rest. But if it could continue bouncing forever, what would be the total the ball traveled?

This is an example of an infinite geometric sequence. Again, each bounce height will be doubled and then the starting height subtracted.

$$2\left(\frac{20}{1 - 0.9}\right) - 20 = 380 \text{ feet}$$

6. There are 88 keys on a piano. The lowest note is an A, whose sound wave has a frequency of 27.5 cycles per second. As we have seen previously in this

book, the frequency of the next note can be found by multiplying by $\sqrt[12]{2}$). Thus, we have the geometric sequence 27.5, 29.1352, 30.8677,... What is the frequency of the highest tone on the piano?

$$a_{88} = 27.5(\sqrt[12]{2})^{88-1} = 4{,}186.01$$

7. What is the decimal 0.6666... written as a fraction? This decimal is an example of an infinite geometric sequence.

$$0.666... = 0.6 + 0.06 + 0.006 + ... = \frac{0.6}{1-0.1} = \frac{2}{3}$$

8. What is 0.123123... written as a fraction?

$$0.123123... = \frac{0.123}{1-0.001} = \frac{41}{333}$$

This method can be used to prove that 0.999... has the same value as 1.

$$0.999... = 0.9 + 0.09 + 0.009 + ... = \frac{0.9}{1-0.1} = \frac{0.9}{0.9} = 1$$

On this point, expect a rousing dissent from the unbelievers.

9. It is suggested that to avoid injury, runners should increase their weekly mileage by no more than 10 percent per week. A runner currently runs 20 miles a week. If she follows this guideline, how many weeks will it take to get to her goal of running 50 miles a week?

$$50 = 20(1.1)^{n-1} \rightarrow \log(2.5) = \log(1.1)^{n-1} \rightarrow n = 10.62 \text{ weeks}$$

So, she shouldn't shoot for a 50 mile week until the eleventh week.

Other Sequences and Series

There are other useful sequences in addition to those that are arithmetic or geometric.

Much work has been done to find better and better approximations for pi. In the 1600s, James Gregory found that $\frac{\pi}{4} = 1 - \frac{1}{3} + \frac{1}{5} - \frac{1}{7} + \frac{1}{9} - \frac{1}{11} + ...$ This works, but takes a long time to converge to pi. A better series was found in the next century by mathematician Leonhard Euler (pronounced "Oiler"). He found that $\frac{\pi^2}{6} = \frac{1}{1} + \frac{1}{4} + \frac{1}{9} + \frac{1}{16} + ...$ Since then, many other series have been found that can be used to compute the value of pi.

The value of e can be found with a series: $e = \frac{1}{0!} + \frac{1}{1!} + \frac{1}{2!} + \frac{1}{3!} + ...$ Many functions can be evaluated with series.

$$\sin(x) = x - \frac{x^3}{3!} + \frac{x^5}{5!} - \frac{x^7}{7!} + ...$$

$$\cos(x) = 1 - \frac{x^2}{2!} + \frac{x^4}{4!} - \frac{x^6}{6!} + \dots$$

$$e^x = 1 + x + \frac{x^2}{2!} + \frac{x^3}{3!} + \dots$$

When trig tables were constructed hundreds of years ago, it was done by evaluating series. An approximate value for sin(2.4) can be found by evaluating $2.4 - \frac{2.4^3}{3!} + \frac{2.4^5}{5!} - \frac{2.4^7}{7!} + \frac{2.4^9}{9!} - \frac{2.4^{11}}{11!}$. Computing this series gives an answer correct to ten-thousandths place. One more term gives an answer correct to ten-millionths place. In a way, finding those values by this method is a thing of the past. But in a way, it isn't. Now we use calculators, but evaluating series is how calculators find those values.[18] What is happening when a student presses buttons on a calculator to find sin(2.4) and gets an answer of 0.675463181? The calculator is likely first changing numbers into binary form, then evaluating the first few terms of $2.4 - \frac{2.4^3}{3!} + \frac{2.4^5}{5!} - \frac{2.4^7}{7!} + \dots$ While it would take mortals a while to do this, the work is being done inside the calculator in a split second.

Calculators can find roots in a variety of ways. To find a value of something like $\sqrt[3]{7}$, the calculator can interpret this as needing to find the x value of f(x) = x³ – 7 = 0. It then uses Newton's Method to solve this equation. This generates a sequence of numbers that converge toward the solution.

Fibonacci was an Italian mathematician of the thirteenth century. The sequence he discovered, and bears his name, begins with two ones. Every two terms are then added to find the next term. Thus we have the sequence 1, 1, 2, 3, 5, 8, 13, 21, 34, 55, 89,... These numbers are found in nature over and over. The spirals of shells, the branching of plants, the seeds on flower heads, and the patterns in pine cones often contain ratios made up of numbers from the Fibonacci sequence.

Circles

Circles, along with ellipses, parabolas, and hyperbolas, are classified as conic sections. All can be found in numerous locations throughout the world. Circles, or in the three-dimensional case—spheres, can be found wherever there is a source of energy being emitted equally in all directions. A rock thrown into a pond creating expanding circles is an example of this.

Coverage Areas

A radio or television station is said to have a certain broadcasting radius, which simply is the radius of the circle of farthest locations that can pick up its signals. Barring obstacles such as mountain ranges, cell phone towers have a certain range. Locations for these towers are chosen to maximize coverage.

Example:

A possible project would be to give the ordered pairs that show locations of cell towers along with their ranges. For example there could be a cell tower located at the point (2,17) with a radius of 8 units. Its equation would be $(x - 2)^2 + (y - 17)^2 = 64$. Information for several towers could be given to students. They then write equations, graph the circles, and estimate the coverage area percentage.

Design

Calculus can be used to find shapes that maximize volumes out of their raw materials. Bubbles are spherical in shape because the filmy substance seeks to form itself to encompass the maximum volume for its surface area.

Just as calculus shows a sphere to be the most efficient shape for the filmy substance, it is true for other materials as well. With no other constraints, a tin can would be constructed with the least amount of metal if it were spherical in shape. Stacking would be an issue, however. So, packaging, buildings, and the homes we live in are probably not going to be spherical. However, the closer to the spherical shape, the more economical they would be. Homes that are elongated lose more heat in winter than those whose length, width, and height are approximately the same.

Developed by architect R. Buckminster Fuller in the 1950s, geodesic domes act much as bubbles do, attempting to make the most efficient use of materials to enclose a space. Technically polyhedral, they are made up of triangles or other polygons, but so many that they appear to be a portion of a sphere sitting on the ground. Again, a sphere would be the most efficient shape for a building, but there would always be the danger of it rolling away.

Lenses

Lenses whose surface is the shape of a slice of a sphere are known as spherical lenses. They are relatively easily and cheaply made. However, light beams passing through spherical lenses will converge close to, but not exactly, to one location. Aspherical (non-spherical) lenses can correct for this distortion. Aspherical lenses can be parabolic, elliptical, or hyperbolic.[19]

Planets and Stars

Planets and stars are formed by a gravitational pull. Because the pull is the same in all directions, we see planets and stars forming spheres rather than some other shape.

Example:

Assuming the point (0,0) to be the middle of the Sun, equations for the surface of the planets could be found.

The radius of the Sun is 432,000 miles. The radius of Mercury is 1,515 miles. The distance from the center of the Sun to the center of Mercury is 36,433,500 miles. Assuming for this example that the Sun and Mercury lie along the x-axis, what are equations for the surfaces of the Sun and of Mercury?

The equation for the surface of the Sun would be $x^2 + y^2 = 432,000^2$.

The equation for the surface of Mercury could be written as $(x - 36,433,500)^2 + y^2 = 1,515^2$.

Ellipses

The equation for an ellipse with a horizontal major axis (which is what will be assumed for these examples) is: $\frac{(x-h)^2}{a^2} + \frac{(y-k)^2}{b^2} = 1$. The foci are c units from the center point (h,k). The value of c can be found with the equation $c^2 = a^2 - b^2$. The eccentricity, $e = \frac{c}{a}$, can be thought of as how flattened the ellipse is. An eccentricity of zero would actually be a circle.

Orbits

Johannes Kepler developed three laws of planetary motion. The first stated that the orbit of any planet is in the shape of an ellipse. The object orbited around, in the case of our solar system, the Sun, is located at one of the two foci of the ellipse. The Moon in its orbit, Halley's Comet, or man-made satellites also have elliptical orbits.

Examples:

1. An interesting project is to find the equation of an orbiting satellite. That can be done with a minimal amount of information. The Moon has an elliptical orbit around the Earth. During 2013, the apogee, when it was farthest away, was 252,183 miles from the Earth. The Moon was 225,025 miles away at its perigee, its closest point. To find the equation of its orbit, use the fact that its apogee has the value a + c and its perigee has the value a − c.

Solving the system: a + c = 252,183 and a − c = 225,025 gives values of a = 238,604 and c = 13,579.

Now, the equation $c^2 = a^2 - b^2$ can be used to find the value of b. Doing so gives b = 238,217. Using a horizontal major axis with the center of at (0,0) yields:

$$\frac{x^2}{252,183^2} + \frac{y^2}{225,025^2} = 1$$

Additionally, it can be found that the Earth would be located at one of the foci (±13,579, 0) and that $e = \frac{13,579}{238,604} = 0.057$.

2. A project could be developed to find the equations of the orbits of

each of the planets, their eccentricities, and a graph of the orbital paths. This could be done in miles or, to reduce the size of the numbers, the calculations can be done in astronomical units. (One AU is the mean distance from the Earth to the Sun—93,000,000 miles.)

3. As seen previously, the Moon's path is nearly circular. This is quite apparent by either graphing the equation or by observing that its eccentricity is close to zero (e = 0.055). Students finding the equations of the planets would see that, while elliptical, the orbits are very close to being circular. It is then interesting to contrast these planetary orbits to those of comets. Halley's Comet, for example, is very elongated. At different points of its orbit, it can be closer to the Sun than Venus and as far away as Pluto. Halley's Comet's farthest distance from the Sun is roughly 35.1 AU and its closest distance is 0.6 AU. Again, a and c can be found by solving a system of two equations. Doing so gives values of a = 17.85 and c = 17.25. With an eccentricity of $\left(\frac{17.25}{17.85} = \right)$ 0.966, Halley's Comet has a far different looking orbit than the planets in our solar system.

4. A planet's apogee and perigee can be found with the equations: Apogee = a(1 + e) and Perigee = a(1 − e), respectively. This can be seen to be true in the previous example involving Halley's Comet.

Apogee = 17.85(1 + .966) = 35.1
Perigee = 17.85(1 − .966) = 0.6

The White House Ellipse

There is a patch of grass between the White House and the Washington Monument known as the Ellipse. It turns out it doesn't just look like an ellipse. It was actually designed to be one. Its major axis has a length of 1,058.26 feet and a minor axis length of 902.85 feet. From this information, the equation for the ellipse can be found, along with the foci and eccentricity. The center of the ellipse nearly had great significance. President Thomas Jefferson wanted the United States to have its own prime meridian going through that point.[20] (Washington, D.C., is filled with geometry, with buildings such as the Octagon and the Pentagon, and numerous triangles, rectangles, and circles in the city plans of architect Pierre L'Enfant.)

Whispering Galleries

Speaking of Washington, D.C., inside the Capitol Building there is what is known as a whispering gallery. It is now named Statuary Hall, but previously was where the House of Representatives met. Public buildings sometimes have rooms in which the ceiling is a partial ellipse (in this book we'll stick with words such as ellipse and parabola, even when, technically, the shape is a three-dimensional ellipsoid or paraboloid.) A room of this shape has the characteristic that a person speaking while standing at a focus point can be heard plainly by a person at the other focus point. Legend has it that a number of important discussions that were meant to be private, were not.

Kidney Stones

Kidney stones can be blasted apart by used of a lithotripter. A lithotripter is basically an ellipse. With this device, doctors can emit ultrasonic waves from a point located at a focus of the ellipse. These ultrasonic shock waves come together at the other focus which is placed at the location of the kidney stone.[21]

Pool Tables

Elliptical pool tables work similarly. Instead of a rectangular pool table, imagine one whose rail is elliptical. There is one hole in the table, which is located at one of the foci. This pool table has the characteristic that any ball hit from the other foci will go into the hole.[22] It will either do so directly or by first banking off a rail.

A good project, short of actually building one, would be to design an elliptical pool table. A student could be given free rein to make any size table, or there could be assigned size requirements that have to be met. An equation could be developed, and a scale version displayed on graph paper, along with the location of the two foci. The designer could also be required to give an equation of the circle representing the hole at the focus.

Parabolas

The ellipse has the property that energy emanating from one focus converges at the other focus. A parabola is similar, although there is only one focus. Parabolas can be of use in both directions. Energy coming in is gathered at the focus. Energy being emitted from the focus is reflected off the parabola, and sent outward.

Searchlights and Flashlights

Flashlights are smaller than searchlights, but other than that, they really are the same thing. Their function is to shine light in a forward direction. To do so, a light source is located at the focus of a parabola. Light from the bulb striking the parabolic surface will be reflected forward in parallel lines. Searchlights work the same. A stronger light source would be used, but would still be located at the focus.

Headlights

Car headlights also work much the same way. Headlights typically make use of a "high-beam" or "low-beam." This effect can be achieved in different ways. One way is to use two different filaments. One is located at the focus, and thus gives a brighter light. The other filament, "low beam," is located at a point other than the focus, and gives a more diffused light.

By the same principle, there are parabolic room heaters for use in homes. The heating unit is located at the focus and the surface of the heater is parabolic.

Parabolic Receivers

As we've seen, parabolas can be used to send out energy. They can also be used to collect energy. Incoming energy will bounce off the parabolic surface and, not surprisingly, congregate at the focus of the parabola. The lighting of the Olympic flame has been accomplished by holding the Olympic torch at the focus of a parabolic mirror, focusing the sun's rays until the torch ignites. Radio telescopes are used by astronomers to collect radio waves from space. Television signals are captured by large satellite dishes or by smaller ones that can be seen attached to many people's homes.

Metallic troughs can be constructed to serve as collectors of solar energy. A reflecting, parabolic metal trough with a receiver tube containing water runs down the middle of the trough (located at the focus) collecting solar energy. The heated fluid becomes steam, travels down the tube, spins a turbine, which leads to a generator, which then produces electricity. These troughs can be modest in size for use by a home, or series of them may be company-owned and cover acres of ground.

Parabolic Microphones

Parabolic microphones are microphones bounded by a parabolic dish which is used to focus the incoming sound. They can be seen on the sideline of football games pointed at the quarterback.[23] Sometimes referred to as "parabs," they are the reason a quarterback calling signals at a game can be heard so clearly during a broadcast.

Suspension Bridges

As seen in the section on hyperbolic functions, the cable on a suspension bridge forms a catenary curve when left solely to the force of gravity. When a roadway is attached, the curve then changes its shape somewhat and becomes a parabola.

Example:

The Golden Gate Bridge's cable hangs from towers that are 500 feet high. The towers are 4,200 feet apart.[24] Assume that at its lowest point, the cable is 10 feet off the ground. If the middle of the bridge is taken to be the origin, then three points of the parabola are (0, 10), (−2,100, 500), and (2,100, 500). This would be enough information to write an equation for the parabola.

Using the form for a parabola, $y = a(x - h)^2 + k$, the vertex and one other point on the curve can be substituted to find the value of a.

$$500 = a(2{,}100 - 0)^2 + 10 \rightarrow a = 0.0001111$$

A formula for the cable on the Golden Gate Bridge could be written as $y = 0.0001111x^2 + 10$

Projectile Motion

The paths of baseballs, footballs, bullets, and missiles are all examples of projectiles. A projectile is such that during its flight, gravity is the only force acting on it.

A projectile's trajectory is always a parabola. In the case of a long jumper, even though arms and legs might be flailing about, the center of gravity of the jumper follows a parabola. A projectile's path could be a very steep curve, as in the case of a baseball pop up that goes a mile high, but horizontally only travels a few feet. Or the parabola could be so flat it might almost appear to be a straight line, such as the path of a 95 mile per hour fastball or a bullet fired from a gun.

If three points of a projectile's path are known, the equation can be found. Two of the points can be fairly easily determined by looking at the beginning and landing point of the projectile. Getting that third point can be a little trickier.

Example:

A volleyball player serves the ball, striking the ball with her hand at a point 6 feet above the court. It barely skims above the net, which is a regulation 7 feet 4 inches off the ground. It strikes the opponent's end line which is 60 feet from the server. What is an equation describing its parabolic trajectory?

Let's begin by taking the server's feet to be at the origin. The ball then starts its flight at the point (0,6). It goes over the net at $(30, 7\frac{1}{3})$, and hits the ground at (60,0). Plugging this information into $y = ax^2 + bx + c$, we obtain three equations.

$$6 = a \cdot 0^2 + b \cdot 0 + c \qquad 7\tfrac{1}{3} = a \cdot 30^2 + b \cdot 30 + c \qquad 0 = a \cdot 60^2 + b \cdot 60 + c$$

With a little work, these can be solved to find a, b, and c. Doing so gives the parabola:

$$y = -0.004814815x^2 + 0.18888889x + 6.$$

If a student has knowledge of differential calculus, a further interesting exploration is to find out how far away, horizontally, the highest point is from the server, and how high the ball goes. [Answer: 19.62 feet away and 7.85 feet high.]

Weightless Flights

Weightlessness can be experienced without going into outer space. During flight, a pilot can cause his aircraft's trajectory to take the shape of a parabola. For roughly twenty to thirty seconds, the plane's passengers will experience weightlessness. Repeated passes through this path can be made to gain additional time. This has been used by NASA for astronaut training. The first plane used for this purpose trained Mercury astronauts and was dubbed the Vomit Comet. The movie *Apollo 13* made these flights to film actors in a state of weightlessness.[25] For a price, this experience is even available for events such as weddings.

Hyperbolas

Sonic Booms

Sonic booms are created when planes travel faster than the speed of sound. In essence, the sound waves produced cannot travel fast enough to get out of its way. A person standing on the ground hears the sound once. However, the boom is not a one-time event, but travels in the wake of the plane. The path of the boom along the ground is in the shape of a hyperbola. The reason for this is clear when we consider why these shapes are called conic sections in the first place. Slicing a cone straight across gives the shape of a circle. The other conics can be seen by slicing a cone at other angles. Slicing downward, parallel to a cone's axis, yields the shape of a hyperbola. The plane, flying at supersonic speeds, creates a cone of sound with the plane at the apex of the cone. If the plane is flying parallel to the ground, any place this cone intersects the ground will form a hyperbola. All along the path of this curve, people are hearing this sonic boom when this cone of sound passes them.

Nuclear Plant Cooling Towers

The hyperbola shape is used for the construction of cooling towers, and are most commonly associated with nuclear power plants. Although this shape has come to be a symbol of nuclear power plants, the cooling towers do not have anything directly to do with generating nuclear power. To prevent overheating, water is used to transfer heat from the reactor. The cooling tower exists simply to cool that water so it can be reused. The cooling comes primarily from air coming in through vents at the bottom of the tower. As it warms, the air rises through the tower. This flow of air past the water helps provide cooling. The hyperbolic shape increases the speed of the draft and, thus, the amount of cooling.[26]

There are other buildings, such as the McDonnell Planetarium in St. Louis, that have been built in the shape of a hyperbola.

Inverse Variation

If a car is going to travel 120 miles, the time it takes for the trip will depend on the rate of travel. The faster the rate; the less time it takes. This relationship can be shown with the equation 120 = rt. Any equation in the form k = xy is an example of inverse variation. Its graph is a hyperbola with a center at the origin and the axes serving as asymptotes. The pressure on a gas and its volume vary inversely. The length of a piano string and its frequency vary inversely. For a given area of a rectangle, possible values for its length and width also vary inversely. There are many other examples.

LORAN Navigation

The LORAN (long range navigation) system was used by boats at sea before the advent of the GPS system. LORAN is still a backup for cases were GPS signals are unavailable or degraded. Recall that the definition of a hyperbola is that it is the set of points that are a fixed difference from two fixed points (the foci). This is the basis of the LORAN system. A network of transmitting stations sends simultaneous signals. The difference of time that a ship at sea received two signals would determine a particular hyperbola the ship was on. This would be compared to the difference between the received signal times from two other transmitting stations. This would determine a second hyperbola. The intersection of these hyperbolas could then be found.[27]

The LORAN system was discontinued by President Obama in 2010, but the U.S. Congress approved the use of a related eLORAN system in 2015.[28]

Telescopes

There are a number of types of telescopes. Refractor telescopes use lenses through which light passes. Reflector telescopes use mirrors. Both types rely on the use of conic sections.[29]

The refractor telescope uses the objective lens (the lens closest to the object being viewed) to collect the light at a focus point. The eyepiece lens then takes that light coming from the focus point and magnifies it. The reflecting telescope does the same, except, rather than going through an objective lens, the light is bounced off a mirror to a focus point, and then magnified by the eyepiece lens.

The mirrors in a reflecting telescope can be spherical, parabolic, or hyperbolic. Each has its disadvantages. Spherical mirrors reflect the incoming light; however, there is not a common focus point for that light. The light hitting near the middle of the mirror is reflected to a different location than light striking the outside part of the mirror. This is called spherical aberration.

A parabolic mirror eliminates this problem. Light entering the telescope parallel to the axis of symmetry is focused, of course, at the focus. However, not all light will enter parallel to the axis and that light will not all collect at the same point. This issue is called coma.

Hyperbolic mirrors eliminate coma. Hyperbolic mirrors work quite well. In fact, because of its precision, a hyperbolic mirror is used in the Hubble space telescope. However, they are very expensive to make.

There are many variations. Telescopes can have primary and secondary mirrors. A Gregorian telescope has a primary mirror that is parabolic, then a secondary mirror that is elliptical. A Cassegrain telescope has a parabolic primary mirror and a hyperbolic secondary mirror. Some telescopes are combination reflector and refractor telescopes. None of them work without the concept of conic sections.

Orbits and Conic Sections

An object's gravitational field has an effect on the path of any nearby object. That path could take the form of any of the conic sections. Or an object coming near another body could have its path altered enough to bend it, but not enough that it enters into a complete orbit around the body. This path would be in the shape of a parabola or hyperbola. The path is determined by the escape velocity of the orbited body. An object approaching at a speed greater than the escape velocity would have its path bent and its path would be a hyperbola. If the velocity of the body happens to be exactly the same as the escape velocity, its path would be a parabola. If the body's velocity is less than the escape velocity, it will fall into a complete orbit around the planet and that orbit will be elliptical.[30]

V

TRIGONOMETRY

Right Triangles

Indirect Measurements

An important use of trigonometry is to measure distances that are difficult, if not impossible, to measure directly. This might include heights of mountains, distances across rivers, or even distances to planets. Here are just a few of many situations where trigonometry can be used to make indirect measurements.

1. At a point 50 feet from the base of a flagpole, the angle of elevation from the ground to the top of the pole is 75 degrees. What is the height of the flagpole?

$$\tan(75°) = \frac{x}{50} \quad [\text{Answer: } x = 186.6 \text{ feet}]$$

2. Suppose you are standing on flat ground, gazing at a mountain. In this case, a right angle exists, but it is buried somewhere in the middle of the mountain. Because of this, none of the sides of the triangle could be measured directly. However, with a little ingenuity, the mountain's height can be found.

You decide to let h be the height of the mountain. You let x be the distance from where you stand to the vertex of that right angle within the mountain. The angle of elevation from where you stand to the mountaintop measures 36.5 degrees. Step back 200 feet and the angle of elevation is now 36.0 degrees. Two equations could be written with this information.

$$\tan(36.5°) = \frac{h}{x} \quad \text{and} \quad \tan(36°) = \frac{h}{x+200}$$

$$h = 0.7400x \text{ and } h = 0.7265(x + 200)$$

Substituting gives 0.74x = 0.7265 + 145.3, which solves to x = 10,763.0 and h = 7,964.6. The mountain has an elevation of 7,965 feet.

Hill Downgrades

Big trucks have a tough time stopping once they get going. It is mostly for their sake that there are signs at the top of mountain passes that might say, "6 percent downgrade next 4 miles." If that is the case, how many feet is the drop in elevation? This would be an excellent opportunity to use the often ignored gradient setting on a calculator. The equation $\sin(6) = \frac{x}{21,120}$ could be solved using gradients, or in degrees, $\sin(5.4°) = \frac{x}{21,120}$. In either case, the the equation solves to show a drop in elevation of 1,987.6 feet.

Television Viewing

THX Limited is an audio/visual company created by George Lucas. THX recommends that the angle from where a person is sitting to each side of a viewing screen be 40° or less.[1]

Example:
Suppose that a television screen measures 50 inches across. To fit the recommendation, what is the closest a person should sit?

Bisecting the 40° angle creates a right triangle. This right triangle would yield the trig equation $\tan(20°) = \frac{25}{x}$. Solving the equation gives a value for x of 68.7 inches or 5 feet 9 inches.

Airport Landing

The standard approach angle for a commercial pilot landing an aircraft is 3 degrees.[2] If a pilot has a current elevation of 2,000 feet, how far in ground distance should the plane be from the airport to attain this angle?

$\tan(3°) = \frac{2,000}{x}$ [Answer: 38,162 feet or 7.2 miles]

Roof Pitch

The pitch of a roof is often written as a ratio of the number of inches it rises for every 12 inches it extends horizontally. The pitch could also be expressed as a number of degrees.

Examples:
1. A 4:12 pitched roof is equal to how many degrees of elevation?

$$\tan(x) = \frac{4}{12} \quad \rightarrow \quad x = 18.4°$$

2. A roof pitch of 35° rises how many inches every foot?

$$\tan(35°) = \frac{x}{12} \quad \rightarrow \quad x = 8.4 \text{ inches.}$$

Track and Field Throwing Events

We have seen previously that the throwing sector for the shot put, discus, and javelin is 34.92 degrees. While some of the reasoning behind the choice of these values was examined in the geometry section, let's take another look at it now that we have trigonometry at our disposal. Trying to accurately lay out a 34.92° angle on a concrete slab seems challenging. The claim by the track rule book is that this angle yields an isosceles triangle whose base is 60 percent of the length of the two congruent sides. Is that so?

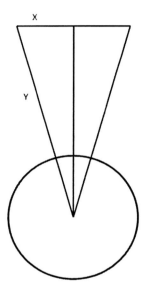

Bisecting this angle yields a right triangle with an acute angle of 17.46°. Suppose it has an opposite side of length x and a hypotenuse of y. Since $\sin(17.46°) = \frac{x}{y}$ we have x = 0.30004y. Remember that this is only half of the triangle. So the side opposite the original angle is 60.008 percent of the two congruent sides. This fact makes marking out the sector fairly easy. For example, going 150 feet from the middle of the shot put circle, then marking off a distance across of 90 feet, would yield the required 34.92° angle.

By a similar process, it can be seen that the 28.96° angle used for the javelin gives another convenient method for setting up a throwing area. Half of 28.96 is 14.48 and sin(14.48°) = 0.25004. So one technique is to get on your hands and knees with a protractor and try to make sure your angle is 28.96°. Instead, use a tape measure and lay out the sector with a proper 0.25 ratio between the opposite side and hypotenuse of a right triangle.

Trigonometry and the shot put.

Guy Wires

The directions for setting up a fifty foot antenna call for guy wires that make an angle of 30 degrees with the ground. How far from the base of the antennae should the guy wires be anchored?

$$\tan(30°) = \frac{50}{x} \quad \rightarrow \quad x = 86.6 \text{ feet}$$

The Leaning Tower of Pisa

The Tower of Pisa, before its adjective "leaning" became necessary, was meant to stand 185 feet tall. It was off-center by 5.5 degrees in the 1980s. Work was done from 1990 through 2001 so the lean is now only 4 degrees.[3] How far from vertical was the top of the tower in the 1980s?

$$\sin(5.5°) = \frac{x}{185}, \text{ so } x = 17.73 \text{ feet.}$$

It is now off-center by 4°. How far from vertical is the top of the tower now?

$$\sin(4°) = \frac{x}{185}, \text{ so } x = 12.905 \text{ feet.}$$

That answer is probably good enough. However, the triangle formed by comparing the original to the new location does not exactly form a right angle. It has angles of 4, 88, and 88 degrees. The Law of Cosines gives an ever so slightly more accurate answer.

The equation $x^2 = 185^2 + 185^2 - 2 \cdot 185 \cdot 185 \cdot \cos(4°)$ gives a value of 12.913 feet—a difference of about one-tenth of an inch.

Air Force Academy Chapel

The chapel at the Air Force Academy in Colorado Springs is an impressive building. It is topped with a row of spires in the form of isosceles triangles. The spires are 150 feet tall with apex angles of 40 degrees. What are the slant heights, the width of the base, and the base angles?

Slant height: $\cos(20°) = \frac{150}{x} \rightarrow x = 159.6$ feet

Width of the base: $\tan(20°) = \frac{y}{150} \rightarrow y = 54.6$. This is half the base, so 109.2 feet.

The base angles are: $180 = 40 + 2x \rightarrow x = 70°$ for each angle.

Oblique Triangles

Distances from a Lighthouse

Two lighthouses are 10 miles apart. With vertex at lighthouse A, the angle extending from lighthouse B to a ship at sea is 27°. With vertex at lighthouse B, the angle from lighthouse A to the ship is 43°. What is the distance from lighthouse A to the ship? From lighthouse B to the ship?

Because all the angles of a triangle must add to 180°, the measure of the angle located at the ship must be 110°. To find the distance from A to the ship, the Law of Sines can be used.

$$\frac{x}{\sin(43°)} = \frac{10}{\sin(110°)} \quad \rightarrow \quad x = 7.26 \text{ miles}$$

Similarly, the distance from lighthouse B to the ship is:

$$\frac{y}{\sin(27°)} = \frac{10}{\sin(110°)} \quad \rightarrow \quad y = 4.83 \text{ miles}$$

Distances Between Ships

Two ships left from the same port. One ship's bearing is N30°E (meaning 30° east of due north) and has traveled 20 miles. A second ship that left at the same time has traveled 14 miles with a bearing of N43°E. How far apart are they?

A sketch of the situation shows a triangle that has two sides of 20 and 14 with an included angle of 13°. The Law of Cosines works for this situation.

$$x^2 = 20^2 + 14^2 - 2 \cdot 20 \cdot 14 \cdot \cos(13°) \rightarrow x^2 = 50.35 \rightarrow x = 7.10 \text{ miles}$$

Infield Dimensions

In major league ballparks, the base paths form a square that is 90 feet on each side. One might assume that the pitcher's mound is in the center of this square. It isn't. It is a little closer to home plate than it is to second base. The distance from the pitcher's mound to home is 60 feet, 6 inches. How far is the throw from the pitcher's mound to first base?

This would be a good problem for which to use the Law of Cosines. The angle with a vertex at home plate and sides going through the mound and first base is 45 degrees. The distances from home plate to the pitcher's mound and first base are 60.5 and 90, respectively. Therefore,

$$x^2 = 60.5^2 + 90^2 - 2 \cdot 60.5 \cdot 90 \cdot \cos(45°) \rightarrow x^2 = 4{,}059.86 \rightarrow x = 63.72 \text{ feet}$$

So, it would be almost four feet farther for a pitcher to throw out a runner at first than it would be to throw to the plate.

Measuring India

It was called the Great Trigonometrical Survey of India. It took place almost two hundred years ago, employing massive, concrete towers that extended from the Indian Ocean, northward across India. These towers served as vertices of giant triangles stretching across the country. Angle measurements were made by what was called the Great Theodolite. A theodolite is basically a telescope mounted on a protractor and is used to measure angles. Theodolites might typically weigh about ten pounds. This one weighed a thousand pounds and took twelve men to carry it. Starting with three towers, they measured the distance between two of them, used the theodolite to find angles, and then used trigonometry to find other distances. Then they moved the theodolite and, using the new distances, solved the adjacent triangle.[4] Thus, they marched across the subcontinent.

They could also use trigonometry to find changes in elevation. Using this process, they found the height of Mount Everest. After a journey of around 2,000 miles they decided Everest had an elevation of 29,002 feet—off by some 30 feet.

Trigonometric Equations

Projectile Motion

Previously in this book, there have been equations relating to an object being thrown vertically into the air or to being dropped. Obviously, objects can move ways other than vertically. How far and how high would a thrown baseball go? It depends both on how fast it is thrown and at what angle. Starting at ground level, a thrown ball would attain a certain maximum height and travel a certain distance before hitting the ground again. The height, h, and distance, d, can be found for a projectile with initial velocity, v (in feet per second), at an angle, θ (in degrees). The acceleration due to gravity is g = 32 ft/sec².

$$h = \frac{v^2}{2g} \cdot \sin^2(\theta) \quad \text{and} \quad d = \frac{v^2}{g} \cdot \sin(2\theta)$$

Additionally, these equations give the location of the object at any particular point in time:

$$x = v \cdot t \cdot \cos\theta \quad \text{and} \quad y = v \cdot t \cdot \sin\theta - \frac{gt^2}{2}$$

One can fairly easily come up with a number of applied problems in connection with these formulas.

Examples:

1. In 2011, Aroldis Chapman of the Cincinnati Reds threw the fastest pitch ever, at 106 miles per hour. At what angle would Chapman throw to get it to the plate at the same level it left his hand?

Even at that speed, he will have to throw upward at a bit of an angle to account for gravity. He wants the ball to travel to the plate, 60.5 feet away. The ball is traveling at 106 mph or 155.467 feet per second.

$$60.5 = \frac{155.467^2}{32} \cdot \sin(2\theta) \quad \rightarrow \quad 0.080 = \sin(2\theta) \quad \rightarrow \quad \theta = 2.3°$$

2. How long does it take to get to the plate?

$$x = v \cdot t \cdot \cos\theta \rightarrow 60.5 = 155.467 \cdot t \cdot \cos(2.3°) \rightarrow t = 0.389 \text{ seconds}$$

3. How high does the ball go above his hand?

Since it will cross home at the same height it left his hand, the highest point would take place when it gets halfway to home plate, or in 0.195 seconds.

$$y = v \cdot t \cdot \sin\theta - 16t^2 \rightarrow y = 155.467 \cdot 0.195 \cdot \sin(2.3°) - 16(0.195)^2 \rightarrow y = 0.608$$

The ball would reach its greatest height at 0.608 feet or a little over 7 inches above his hand.

4. As a project, a student could stand at the goal line of the football field, do his or her best to throw at 45 degrees and note the yard line it lands on. Let us suppose the ball went 100 feet. What information could students find with this little bit of information?

How fast the ball was thrown:

$$d = \frac{v^2}{g} \cdot \sin(2\theta) \quad \rightarrow \quad 100 = \frac{v^2}{32} \cdot \sin(2 \cdot 45°) \quad \rightarrow \quad v = 56.67 \text{ ft/sec}$$

Parametric Equations:

$$x = v \cdot t \cdot \cos\theta \rightarrow x = 56.67 \cdot t \cdot \cos(45°) \rightarrow x = 40.0t$$

$$y = v \cdot t \cdot \sin\theta - \frac{gt^2}{2} \quad \rightarrow \quad y = 56.67 \cdot t \cdot \sin(45°) - \frac{32t^2}{2} \quad \rightarrow \quad y = 40t - 16t^2$$

Maximum height:

$$h = \frac{v^2}{2g} \cdot \sin^2(\theta) \quad \rightarrow \quad h = \frac{56.57^2}{2 \cdot 32} \cdot \sin^2(45°) \quad \rightarrow \quad h = 25.0 \text{ feet}$$

Additionally, students could find how these answers change if, for example, the throwing took place on the Moon (g = 5.3).

Sundials

These days, sundials are more of a novelty time keeping device, but it is interesting to find how they are constructed. There are several techniques used in building them. Here is one.[5]

Begin by drawing a horizontal line segment with the ends corresponding to 6 a.m. and 6 p.m. Another line segment is drawn perpendicular to this segment stretching from the midpoint to a point labeled "12:00 noon."

The triangle that sits atop the sundial is called a gnomon (pronounced NOH-mon). Make one. The gnomon will be a right triangle containing an angle (A) whose measure is the same as the latitude of where it the sundial will be located. The gnomon will be put on the sundial with its right angle on the intersection of the lines and A on the line representing noon.

Now each hour must be found and labeled. The number of hours from either side of noon will be t, and its corresponding number of degrees from the gnomon will be D. These variables are related by the formula, $\tan(D) = \tan(15° \cdot t) \cdot \sin(m \angle A)$.

Examples:

1. A sundial is going to be placed at a latitude of 42°N. Find the angles that correspond to 10:00 a.m. and 2 p.m. (Both are two hours from noon.)

$$\tan(D) = \tan(15° \cdot 2) \cdot \sin(42°) \rightarrow D = 21.1°$$

2. Find the angle that corresponds to 11:00 a.m. [Answer: 10.2°]

3. An archeologist discovers a sundial that was found washed up on shore. He notices that the location of 1:00 p.m. is at a 9° angle. At what latitude was the sundial originally placed? [Answer: 36.2°]

Illuminance

The amount of illumination on a surface is dependent upon the amount of light coming in, at what distance, and at what angle. These variables are related with the formula,[6] $E = \frac{I\cos\theta}{r^2}$, with the illuminance, E, measured in lux; the light coming in is I, measured in candelas; and r is the distance in meters. If the light is shining straight onto the surface, the angle is zero degrees and the formula becomes $E = \frac{I}{r^2}$.

For example, the formula would show that light from a lamp shining on a student's paper at 45° provides 71 percent as much illumination as being straight on. This effect also demonstrates the reason for the different seasons. During the summer months, the Sun is much more directly overhead than during the winter months.

Distances on the Earth

The shortest distance between any two points is a straight line. However, a flat map is not an honest representation of the surface of the Earth. The shortest distance from Nome, Alaska to Moscow, Russia is a path over the Arctic. It would not appear that way on a flat map, but it is quite clear if looking at a globe. A great circle is defined as a circle on a sphere whose center is the same as that of the sphere. It is the largest possible circle that can be drawn on the sphere. An interesting fact is that a great circle containing two points happens to be the shortest distance between those two points. That distance can be found by substituting the radian values of longitude and latitude of each of the two points. If those two points have latitudes and longitudes of x_1, y_1 and x_2, y_2, the following equation gives the shortest distance between them.[7]

$$D = 3963 \cdot \arccos[\sin(x_1)\sin(x_2) + \cos(x_1)\cos(x_2)\cos(|y_2 - y_1|)]$$

Example:

Boise, Idaho, is located approximately at 43.6°W, 116.2°N. Converting to radians gives $x_1 = 0.7610$ and $y_1 = 2.0281$, respectively. Miami, Florida, is located approximately at 25.8°W, 80.2°N. Converting to radians gives $x_2 = 0.4503$ and $y_2 = 1.400$.

Substituting those values into the formula gives a distance between the cities of 2,361 miles.

Speed of Rotation on Earth

Elsewhere, we have found the speed of the Earth as it spins on its axis. However, that is only if you are standing on the Equator. If farther north or south, a person will not travel as far, and so, not as fast. At the North or South Pole a person wouldn't be moving any distance at all. To find the speed, the circumference of the circle for a given latitude must be found and then divided by the time.

The radius of the Earth is approximately 3,960 miles. If x is the radius of a circle of latitude θ, then $\cos\theta = \frac{x}{3,960}$. Therefore, at that latitude, the circle has a radius of x = 3,960$\cos\theta$. Adjusting the formula to find the circumference of the Earth, would give a distance d = 2π(3,960$\cos\theta$).

Finding the time is easy. A rotation happens in one day, which is 24 hours. The speed would be the distance divided by time, so the speed of rotation is $\frac{2\pi(3960\cos\theta)}{24} = 330\pi\cos\theta$.

Examples:
1. How fast is a person moving when standing on the equator? d = 330πcos(0°) = 1,036.7 mph
2. How fast is a person moving when at a latitude of 80°? [Answer: 180.0 mph]
3. What latitude has a rotational speed of 200 miles per hour? [Answer: 78.9°N or 78.9°S]

Cosine Effect Error

As mentioned previously, the police use radar coupled with the Doppler effect to determine the speed of an automobile. The speed is accurate for cars traveling directly toward or away from the officer. However, if a car is traveling at an angle, there is something called the cosine effect error. The true velocity of the automobile can be found with this equation[8]:

$r = v \cdot \cos(\theta)$,   r is the recorded velocity on the radar gun
v is the car's actual speed
θ is angle measure (θ = 0° if car moves directly toward or away.)

This is the simplest case of cosine effect error. It gets more complicated if the cars are also at different elevations or the car is going around a curve.

Examples:
1. You are speeding along at 80 miles per hour. If you are at an angle of 15° off-line from the officer, what would the radar gun read?

$$r = 80 \cdot \cos(15°) \rightarrow x = 77.3 \text{ mph}$$

2. Your speedometer says you were traveling at 43 mph. The officer claims he had you doing 40 mph. That could be explained by being at an angle of how many degrees? [Answer: 21.5°]

3. Police record you going 63 mph at an angle of 10°. What was your actual speed? [Answer: 64.0 mph]

Since it is always true that $\cos(\theta) \leq 1$, the instrument's recorded speed would always be read as being at, or slower than, what the car was really going. Therefore, introducing the cosine effect is not a helpful strategy to avoid paying a traffic ticket. This strategy has actually been tried in court (unsuccessfully).

Snell's Law

The bending of a ray of light as it goes from one medium to another is known as refraction. This is quite apparent when looking at a straw that is partially submerged in a glass of water. Substances have what is known as an index of refraction, n. Air has an index of refraction of 1.0003, ice is 1.31, water is 1.333, and diamond is 2.417. The amount of bending that takes place is based on a measure of two angles. One is called the angle of incidence, θ_1, and is formed by the light ray and the line perpendicular to the dividing line between the two mediums. Entering the other medium, the light will slow or speed up, changing its path. The other angle, the angle of refraction, θ_2, is the angle formed in the other medium by the ray of light and the line perpendicular to the line between the two mediums. Snell's law can be written as: $n_1 \cdot \sin(\theta_1) = n_2 \cdot \sin(\theta_2)$.

Example:
The angle of incidence as light travels from air to ice is 20 degrees. What is the angle of refraction?

$$1.003 \cdot \sin(20°) = 1.31 \cdot \sin(x) \rightarrow x = 15.2°$$

Geometric Shapes

Areas and perimeters are typically computed using lengths of sides, but can also be found with angle measurements and trigonometry.

Examples:
1. Perimeter of a trapezoid with bases of a and b, height h, and base angles of α and β:

$$\text{Perimeter} = a + b + h(\csc\alpha + \csc\beta)$$

2. Area of a regular n-sided polygon with sides length s:

$$\text{Area} = \frac{1}{4} \cdot n \cdot s^2 \cot\left(\frac{180}{n}\right)$$

3. Area of a quadrilateral with diagonals a and b that cross at angle α (either angle at the intersection point will give the same answer):

$$\text{Area} = \frac{1}{2} \cdot a \cdot b \cdot \sin\alpha$$

Satellite Coverage

A satellite orbiting the Earth can't communicate with anyone on the other side of the Earth. It doesn't even cover 50 percent of the Earth. The farthest extent of communication would be marked by the two lines drawn from the satellite that are tangent to the Earth.

An arc of how many degrees on the Earth's surface can be communicated with at a given altitude? The answer can be found with the equation: $\cos(0.5\theta) = \frac{3,960}{3,960 + x}$, where θ is measure of the arc between the two points of tangency and x is the elevation of the satellite.

We can see why this is true by forming a triangle with vertices at one of the points of tangency, the center of the Earth, and the satellite. One of the angles of this triangle happens to be an angle at the Earth's center that measures half of the arc we are looking for (0.5θ). From this angle, the length of the adjacent side is the radius of the Earth (3,960 miles). The length of the hypotenuse is the radius of the Earth plus the elevation of the satellite, 3,960 + x. Since cosine of the angle is the length of the adjacent side over the length of the hypotenuse, we have our equation.

Example:
A satellite is orbiting 20,000 miles above the Earth. How much of the Earth can be seen, and communicated with, from the satellite?

$$\cos(0.5\theta) = \frac{3,960}{3,960 + 20,000} \qquad \rightarrow \qquad \theta = 2 \cdot \cos^{-1}(0.1654) = 161.0°.$$

Running the Curve

A sprinter would want to stay upright while running the straightaway, but rounding the curve, the runner would lean in toward the infield. The optimal amount of lean can be found with this formula[9]:

$\tan(\theta) = \frac{v^2}{gr}$, v is velocity in m/sec, r is the track radius, and g = 9.8 m/sec

The number of degrees of lean from vertical, θ, is needed to balance the horizontal and vertical forces acting on the runner. The horizontal force is mass times acceleration, which, when involving circular motion, is found with m · v^2/r. The vertical force is found with: mg. The amount of lean is then found

with $\tan(\theta) = \frac{\text{horizontal force}}{\text{vertical force}} = \frac{m \cdot \frac{v^2}{r}}{mg} = \frac{v^2}{gr}.$

Example:
Usain Bolt has been clocked with a top speed of 12.27 meters per second. If the radius of the track is 35 meters, what should be his amount of lean? [Answer: 23.7°]

Mercator Maps

During the sixteenth century, Gerardus Mercator created what came to be known as the Mercator map. It takes the points on the Earth and places them on a much more convenient two-dimensional flat map form. Imagine a large piece of paper forming a cylinder around a globe, tangent to the globe at the equator. Mercator found a way to project the points from the globe onto the cylinder.

There are advantages to using Mercator maps. However, a major disadvantage is the distortion that takes place the farther one gets from the equator. Greenland looks the size of Africa, when in reality, Greenland's area is less than 10 percent of Africa's.

Lines of longitude are equally spaced on the map. To transfer a line of latitude onto a y-axis value on the map may be accomplished with the formula[10]:

$$y = r \cdot \ln\left(\tan\left(45° + \frac{\theta}{2}\right)\right), \text{ r is the radius of the Earth and } \theta \text{ is the latitude.}$$

Trigonometric Identities

In many situations, a trigonometric expression can be substituted for another, simplifying the overall expression or placing it in a more convenient form. Trig identities come in many different forms. They might be reciprocal, sum, difference, double, or half-angle identities. Following are a few examples where identities might be used to write an expression in an alternate form.

Range of a Projectile

The horizontal distance covered by a projectile that is propelled with a launch velocity of v, at an angle of θ, is $D = \frac{v^2}{g} \cdot \sin(2\theta)$.

Index of Refraction

The index of refraction, n, is a value that measures the amount of refraction that takes place when light goes from one medium to another. The following formula relates the index of refraction with angles α and β formed by light going through a prism.

$$n = \frac{\sin\left(\frac{\alpha + \beta}{2}\right)}{\sin\left(\frac{\beta}{2}\right)}$$

Mach Number

The Mach number, M, of a plane is the ratio of a plane's speed to the speed of sound. If a plane if flying at Mach 2, it is flying at twice the speed of sound. Similar to the wake behind a boat, a cone is formed with the apex being located at the tip of the plane. The angle of this cone, a, is found with: $\sin\left(\frac{a}{2}\right) = \frac{1}{M}$.

Derivatives

The derivative of cos(x) is –sin(x). The proof of which is found with an addition identity. The definition is based on the familiar formula for slope. The proof would begin like this: If f(x) = cos(x),

then $f'(x) = \frac{\cos(x + \Delta x) - \cos(x)}{\Delta x} = \frac{(\cos(x)\cos(\Delta x) - \sin(x)\sin(\Delta x)) - \cos(x)}{\Delta x} = \ldots$

While this leaves out a good deal of the proof, it does show the use for this identity. Addition identities are also needed to find the derivative of the sine, tangent, cotangent, cosecant, and cotangent functions.

Phase Shifts

With even a quick look at the trig graphs, it can be seen that several are quite similar. For example, the sine graph seems to be just shifted 90° from the cosine graph. That can be proven with an identity. For example, cos(x – 90°) = sin(x), because:

$$\cos(x - 90°) = \cos(x)\cos(90°) + \sin(x)\sin(90°) = \cos(x) \cdot 0 + \sin(x) \cdot 1 = \sin(x)$$

Some make use of more than one identity.

$$\tan(x + 180°) = \frac{\sin(x + 180°)}{\cos(x + 180°)} = \frac{\sin(x)\cos(180°) + \cos(x)\sin(180°)}{\cos(x)\cos(180°) - \sin(x)\sin(180°)} = \frac{-\sin(x)}{-\cos(x)} = \tan(x)$$

Trigonometric Graphs

Trig functions are periodic. It turns out that a great number of natural phenomena are periodic. In a year's time, the Earth takes a path around the Sun and then does it again year after year. The Earth rotates on its axis every day. These happenings cause tides, seasons, average temperatures, and hours of daylight to repeat in predictable ways. Graphs of music sound waves are periodic. The graphs of systolic and diastolic blood pressure are periodic. Visible light, x-rays, television waves, radio waves and more are all part of the electromagnetic spectrum, and can all be represented by trigonometric functions. Just by its name alone, one would suspect that alternating current would be periodic in nature.

While all six trig functions are periodic and have associated graphs, for the sake of simplicity, let us limit ourselves to the sine function. There are three primary ways the sine function can be altered—by changing *its* amplitude, period, and phase shift. The sine function may be written as $y = A\sin(k\theta + c)$. The amplitude of the graph is $|A|$, its period is $\frac{2\pi}{k}$ and its phase shift is $-\frac{c}{k}$. Also, the period is the reciprocal of the frequency and vice-versa. Changes in the variables A, k, c affect the shape of the graph in these areas and the phenomena the graph represents.

Example:

A trig equation can be written that computes the amount of daylight for each day of the year. It can be done with a minimal amount of information.

Some basic information includes: A year has 365 days; the least amount of daylight is on the first day of winter, around December 21; the most is around June 21; and there will be even amounts of daylight and darkness on the equinox dates—March 21 and September 21. Suppose that on that first day of summer there are 3 extra hours of daylight. If that were the case, the longest day of the year would have 15 hours of daylight and the shortest would have 9.

This should be enough information to write a function in $f(x) = A\sin(kx + c)$ form. The amplitude would be equal to 3, so A = 3.

Since the period is $\frac{2\pi}{k}$ and the daylight cycle is 365 days, $365 = \frac{2\pi}{k}$. Therefore, $k = \frac{2\pi}{365}$ or 0.0172.

If the spring equinox fell on the first of January, there would be no phase shift. Instead, it falls about 80 days later. The phase shift is $-\frac{c}{k}$, so $80 = -\frac{c}{0.0172}$. This gives a value of c = –1.376.

The function can now be written as $f(x) = 3\sin(0.0172x - 1.376)$. This equation gives the amount of sunlight above or below the amount on the equinox dates. The function could easily be adapted for it to compute the full amount of daylight by tacking on a "+12": $g(x) = 3\sin(0.0172x - 1.376) + 12$. The x represents the day of the year. For example, for January 10, x = 10. Since January is in the middle of the winter, it is close to the shortest day of the year. In fact, the function gives g(10) = 9.2 hours of daylight. The longest day should be around the 170th day of the year, and this function has g(170) = 15.0, just as it should be.

Sound Waves

Pitch and Loudness

A change in the loudness of a sound corresponds to a change in the amplitude of its associated trigonometric function. Louder sounds would have a greater amplitude. Softer sounds have a smaller amplitude.

A change in the pitch a sound makes is related to a change in the frequency or period of the wave. The note known as middle C has a frequency of 440 cycles per second. This could also be stated that for every second that goes by, middle C goes through 440 wavelength cycles. If its frequency is 440 cycles per second, its wavelength is $\frac{1}{440}$ second per cycle. The value of k can be found with the equation $\frac{2\pi}{k} = \frac{1}{440}$. This gives a value of k = 880π. The trig function y = sin(880πx) would represent the sound made by middle C.

Noise Canceling Headphones

One might think that noise canceling headphones simply muffle unwanted sound by having extra padding to block it out. It is a little more high-tech than that. The headphones analyze the noise and its associated wave, and produces a wave of its own that cancels it. It would seem that adding two sounds together would simply give a louder sound. Often it does, but not necessarily. If the crests and valleys of two waves are mirror opposites of each other, they will cancel each other out. Noise cancelling headphones rely on this concept to get rid of unwanted noise. Mathematically, the characteristics of the trig function representing the unwanted outside noise; its frequency, amplitude, and phase shift; are determined and then a sound is produced whose trig equation is the opposite.[11]

For example, suppose an unwanted sound has a wave formula of y = sin(x). Now, the sound with an equation of y = sin(x – π) could be produced. If the two graphs are combined:

$$\sin(x) + [\sin(x - \pi)] = \sin(x) + [\sin(x) \cdot \cos(\pi) - \cos(x) \cdot \sin(\pi)]$$
$$= \sin(x) + [six(x) \cdot (-1) - \cos(x) \cdot 0]$$
$$= \sin(x) + - \sin(x)$$
$$= 0$$

The equations in most situations would be more complex, but this is, in essence, how noise cancelling head phones work. This same technique is also used with automobile mufflers.[12]

Harmonics

Harmonics are an important part of music. A trumpet and an oboe playing the same pitch at the same volume will still sound quite different. This is because the two instruments will have a different set of harmonics. The frequency of a particular note is known as the fundamental or first harmonic. Multiples of that frequency may be present as well. The second harmonic has twice the frequency of the fundamental tone. The third harmonic has three times the frequency, and so on. The relative strengths of those harmonics are what give instruments their unique sounds.[13]

The fundamental tone is in the form y = Asin(2πft). (Since frequency and period are reciprocals, the proportion $\frac{1}{f} = \frac{2\pi}{k}$ means the value of k is equal to $2\pi f$.) A trumpet with a fundamental frequency of 110 cycles per second, together with its harmonics, might have an associated equation that looks something like this:

$$y = 0.036\sin(220\pi t) + 0.024\sin(440\pi t) + 0.016\sin(880\pi t) + 0.012\sin(1760\pi t) + \ldots$$

Note that the frequency is doubling for each term. The frequencies of sounds, in general, do not have to follow a regular pattern like this, but that is what differentiates sounds that are music and sounds that are just noise. Also, note that the amplitudes are decreasing. Because of this, the fundamental tone will be the main one the listener hears. The softer harmonics influence the quality of the sound produced, and are what give instruments their distinctive sounds. If a different instrument played the same note, the frequencies would be the same, but the values of the amplitudes of the harmonics would be different, creating a different sound quality.

Sound can be converted to an electrical signal and viewed on an oscilloscope. The graph of the previous equation would be a little messy. It would be periodic, but with a lot more hills and valleys than students might be used to seeing. If given that graph, could it be broken apart into its individual terms to find the original harmonics that make it up? Yes. In the early 1800s, French mathematician Joseph Fourier (pronounced "Foureeyay") was researching heat equations. He found that any periodic function could be broken into a series of trigonometric functions. This turned out to be not just applicable to heat, but also to sound analysis and many other areas.

Because the harmonics are decreasing in amplitude, that is, volume, those later harmonics will have less and less effect on the sound. At some point in the series, they couldn't even be heard; either because the amplitudes are so small or the frequencies get to a point that they are above our range of hearing. The amount of space on a hard drive is a valuable commodity. One thing compression software does is to save space on music files by eliminating terms of those later harmonics that take up space, but add little to the sound.

If a person could come up with the correct harmonics, they could duplicate the sound of a flute, trumpet, or really any instrument. In fact, that is what Robert Moog did in 1960s when he developed what would come to be called the Moog Synthesizer. (Moog is a Dutch name and is pronounced similarly to the word "vogue.") The word "synthesize" means to put together. In terms of music, a synthesizer puts various sounds together to act as harmonics. Doing this creates different sounds which mimic all different kinds of instruments.

The Electromagnetic Spectrum

Michael Faraday discovered that magnetism and electricity were connected forces. James Clerk Maxwell discovered the electromagnetic spectrum and a set of equations that describe its workings. He found that these electromagnetic waves travel at the same speed: 186,000 miles per second or 3×10^8 meters per second. Light, radio waves, television waves, gamma rays, and x-rays are basically all the same electromagnetic wave. These waves have associated trigonometric functions. The fact that they have differing frequencies is what distinguishes them from one another.

For example, in the visible light portion of the spectrum, a wave has a particular frequency that our eyes perceive as the color yellow. If that frequency of that light was a little higher, we would perceive it as the color green. If the frequency of the green light was made one thousand times greater, we could not see it. It would be an x-ray.

The electromagnetic spectrum can be divided into sections—radio waves, microwaves, infrared, visible light, ultraviolet, x-rays, and gamma rays. They are typically classified by frequency as measured in hertz (Hz) or cycles per second. The dividing lines between these sections are not precise, but are helpful for classification sake.

Electromagnetic Wave	*Frequency (in hertz)*
Radio and Television	less than 3×10^9
Microwaves	3×10^9 to 3×10^{12}
Infrared	3×10^{12} to 4×10^{14}
Light	4×10^{14} to 8×10^{14}
Ultraviolet	8×10^{14} to 3×10^{17}
X-Rays	3×10^{17} to 3×10^{19}
Gamma Rays	greater than 3×10^{19}

Light

The category of light could be divided up even more. Isaac Newton showed that light going through a prism can be separated. Newton arbitrarily came up with the colors red, orange, yellow, green, blue, indigo, and violet to describe this spectrum. Anyone who ever had the big box of Crayola crayons knows there are a lot more than seven colors. There are a lot of numbers between the frequencies of 4×10^{14} and 8×10^{14}, and as a result, a lot of colors. For example, the general region of what we might identify as being different shades of yellow consists of frequencies from 5.1×10^{14} to 5.3×10^{14}. Likewise, the general region of shades of green is from 5.3×10^{14} to 6.1×10^{14}.

The phenomenon known as red shift has to do with the frequency of light. Sound will increase or decrease in pitch as it comes toward or goes away from a person. Relative to that person, the sound waves become more compacted or more spread out. This is known as the Doppler effect. The same is true with light. Scientists feel the universe is expanding because light from stars are shifted toward the red end of the light spectrum. This implies the stars' frequencies are lower because they are moving away from us.

Broadcast Frequencies

Another area where frequency is important is at the lower end of the spectrum. The feature that distinguishes radio stations or television stations is their broadcast frequency. The frequency of a radio station is that associated with its electromagnetic wave. Suppose you are listening to the radio whose display says it is on 1150 AM. The "AM" stands for Amplitude Modulation. The wave must be modulated in some way to encode information on it. When the signal is transmitted, it is done by a carrier wave whose frequency and phase stays the same, but whose amplitude varies. If you look closely, the letters kHz are probably printed somewhere on the radio dial. This corresponds to 1150 kilohertz = 1150×10^3 hertz = 1,150,000 cycles per second.

Suppose the stereo is switched to 99.5 MHz FM. This is 99.5 Megahertz = 99.5×10^6 hertz = 99,500,000 cycles per second. Here, the FM refers to Frequency Modulation. In this case, the signal is modulated by changing the frequency. The 99.5 MHz is known as the resting frequency, but with slight variations during the transmission.

On a radio dial, AM stations are generally shown as being between 530 and 1710 kHz (5.3×10^5 to 1.7×10^6 hertz). The FM stations are shown as being between 87.7 and 107.9 MHz (8.88×10^7 to 1.079×10^8 hertz). The various television stations we watch are also based on their frequencies. Television has sections of the spectrum that include VHF stations, standing for the appropriately named "very high frequency," and UHF stations, standing for "ultra high frequency."

Wavelength

The wavelength is the distance from one crest to another, and is inversely proportional to the frequency. Gamma rays have a higher frequency than ultraviolet radiation. This also means that the wavelength of gamma rays is shorter than radio waves.

Example:

The wavelength can be found by dividing the speed of electromagnetic radiation by its frequency. Radio has the lowest frequency, so it has a longer wavelength than the other forms of radiation. Shortwave radio is used in communication for planes or boats at sea. Shortwave radio is called that because, at the time of its development in the 1920s, it was a shorter wavelength than the other radio signals being broadcast. Two megahertz is a possible frequency for shortwave radio.

$$\text{wavelength} = \frac{3.0 \times 10^8}{2.0 \times 10^6} = 1.5 \times 10^2 = 150 \text{ meters}$$

Amplitude

In the ocean, stronger storms produce higher waves. Not just in the medium of water, but generally, a wave's power is related to its amplitude. A change in the amplitude of a wave coming from a radio station would change its power, and thus, its broadcasting radius. Many microwave ovens have low, medium, and high settings. Lasers can drill through concrete walls and can be used for eye surgery. You would not want your eye doctor to get his settings confused. For each of these, the amplitude of the electromagnetic wave relates to its power.

Blocking, Absorbing and Passing Through

The particular frequency/wavelength of a wave can cause it to be absorbed by a certain object, or it might bounce off that object, or just pass right through. Generally, waves of low frequency also have low energy and are more easily blocked.

Examples:

1. Cell phones, traffic speed cameras, wireless data, and radar make use of microwaves. The way most people think of microwaves is their use in heating foods. Microwaves have a frequency such that they are absorbed by water molecules, causing them to heat up. This is why food, but not the plate it sits on, will heat up when put into a microwave oven. The walls of the microwave oven then block any microwaves from getting out.

2. Between microwaves and visible light is the infrared section. Infrared radiation can send signals to our televisions via remote control. Without it, we would have to actually get up and walk to the television to change the channel. They also allow us to "see" in the dark and make it possible for alarm systems to detect intruders. The primary source of infrared radiation is heat. Anything producing heat, such as humans, can be located even when visible light is not available. Devices such as night vision binoculars work by being able to pick up this infrared radiation. Infrared radiation is used to study the universe. Because of its frequency, it can pass through dense gas clouds when visible light cannot.

3. Next up on the spectrum is the range of visible light. When we see certain colors, such as red, it is because red is reflected off that surface while all of the other colors are absorbed by it.

4. Beside producing visible light, the Sun also produces great amounts of ultraviolet radiation. It is this UV radiation, and not the visible light, that causes suntans and sunburns. For those that like such things, tanning beds make use of ultraviolet light to darken the skin. A substance known as

melanin in our skin absorbs UV radiation. This causes the skin to darken, giving the tan. When there is more UV radiation than the melanin can handle, sunburn and assorted cellular damage can result. Luckily, most UV radiation is blocked by the Earth's atmosphere. Specifically, it is blocked to a great degree by ozone in the atmosphere. This is the reason that the ozone hole is of such concern to scientists. Another level of protection is sunscreen, which contains some chemicals that absorb and some chemicals that reflect UV rays, not letting them reach the skin.

5. X-rays possess an even higher frequency, and thus more energy. Their frequency is between $3x10^{16}$ and $3x10^{19}$ cycles per second. They have enough energy to penetrate our skin and the organs of our body, but not quite enough energy to penetrate bone. This is what makes x-rays so useful for finding out if you actually did break your arm playing basketball or it just really hurts.

6. The highest frequencies, and therefore possessing the most energy, belong to the gamma rays. They have energy to the point of being dangerous. They can come to us from space or from nuclear reactions. Unlike x-rays, they don't just penetrate human tissue; they destroy it. Gamma rays do have positive uses. They are often used to sterilize equipment. The FDA allows gamma radiation of food to destroy harmful bacteria. Gamma rays can be used to destroy cancer cells. Because they have so much energy, this radiation therapy can also have damaging effects on other parts of the body. This results in many of the negative side effects associated with this therapy.

Lasers

Lasers are usually associated with visible light, but could be applied to other parts of the electromagnetic spectrum as well. Lasers are created by introducing energy which causes atoms to have higher-energy electrons. As those electrons return to their previous energy levels, particles of light, or photons, are released. If this is controlled in a certain way, laser light is produced. Lasers are useful in a number of applications including laser eye surgery, bar code scanners, CD players, and laser pointers.

Lasers have characteristics different from regular light. Rather than a mixture of all the colors, lasers are always one specific color. The energy levels, thus the frequency/color, will be the same. Rock concerts may have combinations of red, yellow, or green lasers. Secondly, "It is organized—each photon moves in step with the others. This means that all of the photons have wave fronts that launch in unison."[14] In other words, the waves have the same amount of phase shift. Finally, visible light usually spreads out in all directions. Laser light is very much directed.

Alternating Current

In 1831, British scientist Michael Faraday discovered that a coil of wire could be rotated between the poles of a magnet, causing electric current to flow through the wire. As this rotation happened between the poles, the electricity alternately would flow in one direction, then the other. Alternating current was born. Direct current is produced by a battery and flows in one direction. It works well in many situations. For big jobs, such as supplying power to an entire city, alternating current is the best source of that power. While direct current employs a battery which uses stored chemical energy, alternating current uses a generator, which works much the same way Faraday's did a couple of hundred years ago.

But what is going to turn this coil? The coil can be turned by water or wind power. Or the mechanical energy of steam can be used. The steam is made from heated water—water being heated by nuclear fission or burning coal.

Trig Graphs of Alternating Current

Because of alternating current's periodic nature, many facets of it can be graphed as trigonometric functions. If the x-axis represents time, the y-axis could represent electrical concepts such as current, voltage, or power. For example, a formula to relate the time (t) in seconds and the current (I) in amperes is $I = A \cdot \sin(kt + c)$.

Examples:

1. A specific equation for current might be $I = 20\sin(120\pi t)$. This equation shows that there is a maximum current of 20 amperes. The period is found by using the formula, period $= \frac{2\pi}{k}$. For this equation, the period is $\frac{2\pi}{120\pi}$, which means it takes 1/60 of a second for one cycle to occur. Since the period and frequency are reciprocals of each other, the current has a frequency of 60 hertz.

2. The alternating current used in the United States operates at 60 cycles per second. Other countries often operate at 50 cycles per second.[15] What is an equation showing current that has a maximum of 30 amps and operating at a frequency of 50 hertz? [Answer: $I = 30\sin(100\pi t)$]

3. The concept of peak-to-peak voltage is often used in electronics. It is the difference between the highest and lowest voltage measures. It can be found by doubling the amplitude.

Leading and Lagging Current

The phase shift is an important concept when using alternating current. Inductors store electrical energy in a magnetic field. In an inductive circuit, current builds up more slowly than voltage. An electrical engineer might say that the current lags the voltage. This lagging would be expressed as a certain number of degrees. There is essentially a phase shift between the current and the voltage.

Capacitors store energy in an electric field. In capacitive circuits, current must flow in order to charge the capacitor plates and raise the voltage. This means that current leads the voltage. For either inductors or capacitors, it could be stated that the current and the voltage are out of phase.

AFTERWORD

However you got here, whether you read straight through, or you were more selective, I hope this book has been helpful for you. There are more applications than those included here. I left some out purposely or because some of the categories in the book already had plenty. Appropriate applications can be hard to find, but they are there.

When I started teaching, I would occasionally hear from students, "Where are we ever going to use this?" I didn't have a good answer. I wouldn't say it, but I often thought, "Where *are* they ever going to use this?" I feel I could give a pretty good answer now. Part of this is due to the research I did for this book. Part is just bits and pieces I've collected as I've gone through life. The section of the book about concert insurance was something I directly experienced. I knew something about the LORAN navigational system from talking to a person using it while he operated his boat in the ocean. I knew the industry standard for airplane descent was three degrees from asking a commercial pilot as I got off a plane a few months ago.

While this book represents the applications I've collected over time, you undoubtedly have encountered your own applications that you can share with your students. In your teacher's edition or in your own notes, you might consider making your own list. You can, like I typically did, simply rely on your memory to fit the application into the correct spot in the school year. It turns out that method doesn't always work. If you are well acquainted with the application, a line or two is probably all that you need.

For additional ideas, I do have a blog in which I try to showcase a weekly application. Some are repeats of ones already included in this book, but most are on new topics. Its address is http://wherearewegoingtousethismath.blog spot.com/.

By and large, it has been an enjoyable process in writing this book. Some parts came easily, as they were examples I had used for years. There were others I knew very little about. Those were tougher. There are certain paragraphs that represent hours of researching and struggling to get the ideas straight in my head. In spite of that, writing this book was a very enjoyable, interesting endeavor.

This book was enjoyable mostly for the same reason we get into teaching in the first place. My favorite education-related quote is by William Butler Yeats. He said, "Education is not the filling of a pail, but the lighting of a fire." Most get into education to light fires; not to fill pails. One reason some students dislike mathematics is that they see it strictly as a collection of facts and techniques and theorems. That is difficult to argue with if there is no fire, or even a few embers. One way to help fan those flames is to show students how useful mathematics really is; maybe to the point that you never again hear, "Where are we ever going to use this?"

CHAPTER NOTES

Chapter I

1. Martha K. Smith, "History of Negative Numbers," M 326K, 19 Feb. 2001, Web, https://www.ma.utexas.edu/users/mks/326K/Negnos.html.

2. "Tolerances for Rectangular Bars," Tolerance Values, Web, http://www.alumeco.com/Knowledge-and-Technique/Tolerances/Tolerances-for-rectangular-bars.aspx.

3. Alison Maxwell, "Supreme Court Rejects Census Sampling Plan," *Government Executive*, 25 Jan. 1999, Web, http://www.govexec.com/federal-news/1999/01/supreme-court-rejects-census-sampling-plan/1597/.

4. "Employment Situation News Release," U.S. Bureau of Labor Statistics, Web, http://www.bls.gov/news.release/empsit.htm.

5. J. Louie, "What Is Richter Magnitude?" 9 Oct. 1996, Web, http://crack.seismo.unr.edu/ftp/pub/louie/class/100/magnitude.html.

6. Jim Kolb, "Water Pressures at Ocean Depths," NOAA Pacific Marine Environmental Laboratory (PMEL), Web, http://www.pmel.noaa.gov/eoi/nemo1998/education/pressure.html.

7. "Chapter 6 Fats," 9 July 2008, Web, http://health.gov/dietaryguidelines/dga2005/document/html/chapter6.htm.

8. "How to Calculate the Energy Available from Foods," TimetoRun Nutrition, Web, http://www.time-to-run.com/nutrition/calculate-energy.htm.

9. "How to Calculate the Energy Available from Foods," TimetoRun Nutrition, Web, http://www.time-to-run.com/nutrition/calculate-energy.htm.

10. Paige Waehner, "Understanding Your Maximum Heart Rate," Verywell, 14 Jan. 2016, Web, http://exercise.about.com/od/healthinjuries/g/maxheartrate.htm.

11. "NCAA and NFL Passing Efficiency Computation," Web, http://football.stassen.com/pass-eff/.

12. "QB Rating Calculator," Web, http://brucey.net/nflab/statistics/qb_rating.html.

13. Barton D. Schmitt, "Acetaminophen (Tylenol) Dosage Table," Kid Care: Acetaminophen Dosage Chart for Children and Infants, Web, http://www.stlouischildrens.org/articles/kidcare/acetaminophen-tylenol-etc-dosage-table.

14. "Saffir-Simpson Hurricane Wind Scale," Web, http://www.nhc.noaa.gov/aboutsshws.php.

15. Toshiba Website, 24 Apr. 2013, Web, http://forums.toshiba.com/t5/Televisions-Knowledge-Base/How-To-Best-distance-and-viewing-angle-for-your-HDTV/ta-p/281923.

16. Dave Roos, "How Computer Animation Works," 28 April 2008, Web, HowStuffWorks.com, http://entertainment.howstuffworks.com/computer-animation.htm.

17. Kim Ann Zimmermann, "Temperature: Facts, History & Definition,"LiveScience, TechMedia Network, 20 Sept. 2013, Web, http://www.livescience.com/39841-temperature.html.

18. Kim Ann Zimmermann, "Temperature: Facts, History & Definition,"LiveScience, TechMedia Network, 20 Sept. 2013, Web, http://www.livescience.com/39841-temperature.html.

19. Margaret Rouse, "What Is X and Y Coordinates? Definition from WhatIs.com," WhatIs.com, Apr. 2005, Web, http://whatis.techtarget.com/definition/x-and-y-coordinates.

20. Sarah Horton and Patrick Lynch, "Color

Displays," GRAPHICS: 2002, Web, http://webstyleguide.com/wsg2/graphics/displays.html.

21. Andy Serkis, "Gollum in The Lord of the Rings: Performance Capture," 2016, Web, http://www.serkis.com/performance-capture-gollum.htm.

22. Ken Skorseth and Ali A. Selim, "Gravel Roads: Maintenance and Design Manual," Washington, D.C.: U.S. Dept. of Transportation, Federal Highway Administration, 2000, Web, https://www.epa.gov/sites/production/files/2015–10/documents/2003_07_24_nps_gravelroads_gravelroads.pdf.

23. ADA Ramp Codes, http://www.guldmann.net/Files/Billeder/GuldmannProdukter/Stepless/Transportable%20ramper/Shared%20files/US/ADA%20Ramp%20Codes_US.pdf.

24. ADA Ramp Codes, http://www.guldmann.net/Files/Billeder/GuldmannProdukter/Stepless/Transportable%20ramper/Shared%20files/US/ADA%20Ramp%20Codes_US.pdf.

25. "Is Your Roof Angle Suitable for Solar Panels?" Which Solar Panel Angle Is Best?, Web, https://www.solarquotes.com.au/panels/angle/.

26. Timothy Schmidt and David Baxter, "The Physics of Scuba Diving," 2003, Web, http://physics.itsbaxter.com/pressure.html.

27. "How to Calculate Boiling Point—Video & Lesson Transcript," Study.com, Web, http://study.com/academy/lesson/how-to-calculate-boiling-point.html#transcriptHeader.

28. "How to Calculate Boiling Point—Video & Lesson Transcript," Study.com, Web, http://study.com/academy/lesson/how-to-calculate-boiling-point.html#transcriptHeader.

29. Anne Marie Helmenstein, "Learn the Chemistry Behind Gasoline and Octane Ratings," About.com Education, 10 Oct. 2015, Web, http://chemistry.about.com/cs/howthingswork/a/aa070401a.htm.

30. Debbie Hadley, "How to Use Crickets to Calculate Temperature," About.com Education, Web, http://insects.about.com/od/insectfolklore/a/crickets-temperature.htm.

31. Juleyka Lantigua-Williams, "Are You in the Shrinking Middle Class?" *The Atlantic*, 9 Dec. 2015, Web, http://www.theatlantic.com/politics/archive/2015/12/are-you-in-the-shrinking-middle-class-take-this-2-step-test/433599/.

32. David McAuley, "Mean Arterial Pressure: MAP Calculator," Web, http://www.globalrph.com/map.cgi.

33. "Televisions Knowledge Base," How To: Best Distance and Viewing Angle for Your HDTV, Web, http://forums.toshiba.com/t5/Televisions-Knowledge-Base/How-To-Best-distance-and-viewing-angle-for-your-HDTV/ta-p/281923.

34. United States National Park Service, "Gateway Arch Fact Sheet," National Parks Service, U.S. Department of the Interior, Web, https://www.nps.gov/jeff/planyourvisit/gateway-arch-fact-sheet.htm.

35. N. Madison and Bronwyn Harris, Wise Geek, Conjecture, 7 Apr. 2016, Web, http://www.wisegeek.com/what-is-supply-and-demand.htm.

36. Marshall Brain and Tom Harris, "How GPS Receivers Work," 25 September 2006, Web, HowStuffWorks.com. http://electronics.howstuffworks.com/gadgets/travel/gps.htm.

Chapter II

1. "A Selection from Symbolic Logic. Lewis Carroll," YesFine.com, Web, http://www.yesfine.com/carroll_symbolic_logic.htm.

2. Alwyn Olivier, "2C," Feb. 1999, Web, http://academic.sun.ac.za/mathed/174/circlesregionschords.pdf.

3. Jon Herring, "Sold to the Highest Bidder: The Fatally Flawed Food Pyramid," Early to Rise, Web, http://www.earlytorise.com/our-no-hassle-plan-for-healthy-living/sold-to-the-highest-bidder-the-fatally-flawed-food-pyramid/.

4. "The Fifth Postulate," 2015, Web, http://www.physicscentral.com/explore/writers/bardi.cfm.

5. Harvard School of Engineering and Applied Sciences—CS 152: Programming Language, Curry-Howard Isomorphism, 2 Apr. 2013, Web, http://www.seas.harvard.edu/courses/cs152/2013sp/lectures/lec17.pdf.

6. B.H. Suits, "Pythagorean Scale," Web, http://www.phy.mtu.edu/~suits/pythagorean.html.

7. "Tuning and Ratio," Nrich.maths.org, Web, https://nrich.maths.org/5453.

8. "Gas Laws," Web, http://chemistry.bd.psu.edu/jircitano/gases.html.

9. "Gas Laws," Web, http://chemistry.bd.psu.edu/jircitano/gases.html.

10. NFL Rulebook, Web, http://operations.

nfl.com/the-rules/2015-nfl-rulebook/#rule-2.-the-ball.

11. Marshall Brain, "How Bicycles Work," 1 April 2000, Web, HowStuffWorks.com, http://adventure.howstuffworks.com/out-door-activities/biking/bicycle.htm.

12. "Throw Ratios and Viewing Distances," Web, http://www.theprojectorpros.com/learn-s-learn-p-theater_throw_ratios.htm.

13. Todd Gray, "Tolkowsky Ideal Cut Diamond Proportions & Tolkowsky Diamond Design," Nice Ice Diamonds Comments. 2015, Web, http://niceice.com/tolkowsky-ideal-cut/.

14. "What Is Fowler's Position? (Definition and Explanation)," www.nursefrontier.com, 25 Jan. 2016, Web, http://www.nursefrontier.com/what-is-fowlers-position-definition-and-explanation/.

15. David P. Stern, "(8) The Round Earth and Christopher Columbus," 31 Mar. 2014, Web, http://www.phy6.org/stargaze/Scolumb.htm.

16. "Eratosthenes," Famous Scientists, famousscientists.org, 21 Jun. 2014, Web, http://www.famousscientists.org/eratosthenes/.

17. David P. Stern, "(9a) May Earth Be Revolving around the Sun?" 2 Apr. 2014, Web, http://www.phy6.org/stargaze/Sarist.htm.

18. "Astronomy 101 Specials: Aristarchus and the Size of the Moon," Web, http://www.eg.bucknell.edu/physics/astronomy/astr101/specials/aristarchus.html.

19. David P. Stern, "(9a) May Earth Be Revolving around the Sun?" 2 Apr. 2014, Web, http://www.phy6.org/stargaze/Sarist.htm.

20. David P. Stern, "(9a) May Earth Be Revolving around the Sun?" 2 Apr. 2014, Web, http://www.phy6.org/stargaze/Sarist.htm.

21. "Laying Out Applications," Pro PHP-GTK: 87–118, Web, http://www.pntf.org/officials/ivars/Throwing_Event_Sector_Angles_Rev_F1.pdf.

22. "Laying Out Applications," Pro PHP-GTK: 87–118, Web, http://www.pntf.org/officials/ivars/Throwing_Event_Sector_Angles_Rev_F1.pdf.

23. Marianne Freiberger, "Maths behind the Rainbow," Plus.maths.org, Web, https://plus.maths.org/content/rainbows.

24. Marianne Freiberger, "Maths behind the Rainbow," Plus.maths.org, Web, https://plus.maths.org/content/rainbows.

25. "What Is Seismology and What Are Seismic Waves?" Web, http://www.geo.mtu.edu/UPSeis/waves.html.

26. "How Do I Locate That Earthquake's Epicenter?" Web, http://www.geo.mtu.edu/UPSeis/locating.html.

27. Anne Marie Helmenstein, "What Are the Types of Crystals?" About.com Education, 2 Dec. 2014, Web, http://chemistry.about.com/cs/growingcrystals/a/aa011104a.htm.

28. Gary E. Kaiser, "Size and Shapes of Viruses," June 2015, Web, http://faculty.ccbcmd.edu/courses/bio141/lecguide/unit4/viruses/ssvir.html.

29. Mark Hachman, "Nvidia Finally Gets Faces Right—Until They Open Their Mouths," ReadWrite, 21 Mar. 2013, Web, http://readwrite.com/2013/03/21/nvidia-finally-gets-faces-right-until-they-open-their-mouths/.

30. Ernest Zebrowski, Jr., *A History of the Circle* (New Brunswick, NJ: Rutgers University Press, 2000), page 7.

31. "10 Amazing Facts About the Geodesic Sphere of Spaceship Earth," DisneyFanatic.com, 11 Sept. 2014, Web, http://www.disneyfanatic.com/10-amazing-facts-geodesic-sphere-spaceship-earth/.

32. Harold W. Kroto, "Fullerene," Encyclopedia Britannica Online. 11 Nov. 2014, Web, http://www.britannica.com/science/fullerene.

33. Tom Benson, "Lift Equation—Activity," 12 June 2014, Web, http://www.grc.nasa.gov/WWW/k-12/BGA/Melissa/lift_equation_act.htm.

34. Tom Benson, "Lift Equation—Activity," 12 June 2014, Web, http://www.grc.nasa.gov/WWW/k-12/BGA/Melissa/lift_equation_act.htm.

35. "Design and Construction of Aquaculture Facilities," *Aquaculture Engineering*: 294–320, Web, http://www.phy6.org/stargaze/Sarist.htm.

Chapter III

1. "Stopping Distance Formula," Web, http://www.softschools.com/formulas/physics/stopping_distance_formula/89/.

2. "Kilobytes Megabytes Gigabytes," Web, https://web.stanford.edu/class/cs101/bits-gigabytes.html.

3. B.H. Suits, "Pythagorean Scale," Web, http://www.phy.mtu.edu/~suits/pythagorean.html.

4. "Tuning and Ratio," Nrich.maths.org, Web, https://nrich.maths.org/5453.

5. M. Erk Durgun, "6. Distance—Time—Relativity," Space-Time Interval, Web, http://www.unitytheory.info/space-time_interval.html.

6. "Pythagorean Expectation Calculator & Formula," Web, http://www.had2know.com/sports/pythagorean-expectation-win-percentage-baseball.html.

7. "Pythagorean Expectation Calculator & Formula," Web, http://www.had2know.com/sports/pythagorean-expectation-win-percentage-baseball.html.

8. Clay Davenport and Keith Woolner, "Baseball Prospectus—Revisiting the Pythagorean Theorem," Baseball Prospectus, 30 June 1999, Web, http://www.baseballprospectus.com/article.php?articleid=342.

9. "Adjusting Football's Pythagorean Theorem," Pro-football-reference.com, Blog, 18 July 2007, Web, http://www.pro-football-reference.com/blog/?p=337.

10. "Kepler's Three Laws," Web, http://www.physicsclassroom.com/class/circles/Lesson-4/Kepler-s-Three-Laws.

11. Chris Impey, "Nuclear Reactions in the Sun," 2012, Web, http://m.teachastronomy.com/astropedia/article/Nuclear-Reactions-in-the-Sun#.

12. "OpenAnesthesia," Web, https://www.openanesthesia.org/poiseuilles_law_iv_fluids/.

13. Simon Hartley, "The BMI Formula," How to Calculate, Web, http://www.whathealth.com/bmi/formula.html.

14. "Earth Atmosphere Model—Imperial Units," Ed. Nancy Hall, Web, http://www.grc.nasa.gov/WWW/k-12/airplane/atmos.html.

15. "Formula Used to Calculate Wind Chill," USATODAY.com, 30 Oct. 2001, Web, http://usatoday30.usatoday.com/weather/winter/windchill/wind-chill-formulas.htm.

16. "Formula Used to Calculate Wind Chill," USATODAY.com, 30 Oct. 2001, Web, http://usatoday30.usatoday.com/weather/winter/windchill/wind-chill-formulas.htm.

17. Vassilios Spathopoulos, "Coefficient of Restitution," In Sports, Web, http://www.topendsports.com/biomechanics/coefficient-of-restitution.htm.

18. Medha Godbole, "Official Basketball Size," Buzzle.com, 5 June 2015, Web, http://www.buzzle.com/articles/official-basketball-size.html.

19. "Welcome to VMAR (Vehicle Mechanical and Accident Reconstruction, LLC)!" June 2011, Web, http://www.vmar.net/email_jun11.htm.

20. Doug Davis, "Circular Motion, (and Other Things)," 2001, Web, http://www.ux1.eiu.edu/~cfadd/1350/06CirMtn/FlatCurve.html.

21. Cars Traveling Around a Banked Curve, Web, http://plaza.obu.edu/corneliusk/up1/bc_f.pdf.

22. Matt Webb Mitovich, "Number of 'Nielsen Families' That Fuel TV Ratings to Grow 'Significantly,'" TVLine Comments, 29 May 2014, Web, http://tvline.com/2014/05/29/tv-ratings-nielsen-to-increase-sample-size/.

23. "How the Government Measures Unemployment," U.S. Bureau of Labor Statistics, 8 Oct. 2015, Web, http://www.bls.gov/cps/cps_htgm.htm.

24. "BMI Calculator," Body Mass Index, Web, http://bmi-calories.com/bmi-calculator.html.

25. "STR: A Brief History of Einstein's Special Theory of Relativity," Web, http://www.twow.net/ObjText/OtkCaLbStrB.htm.

26. Mike Wall, "The Most Extreme Human Spaceflight Records," Space.com, Web, http://www.space.com/11337-human-spaceflight-records-50th-anniversary.html.

27. "Single-Season Leaders & Records for Power-Speed," Baseball-Reference.com, Web, http://www.baseball-reference.com/leaders/power_speed_number_season.shtml.

28. Tom Harris, "How Cameras Work," 21 March 2001, Web, HowStuffWorks.com, http://electronics.howstuffworks.com/camera.htm

29. "Lens Maker's Formula," Tutorvista.com, Web, http://www.tutorvista.com/content/physics/physics-iv/optics/lens-makers-formula.php.

30. "Volume/Time—IV Drop Rate Questions," DosageHelp.com, Web, http://www.dosagehelp.com/iv_rate_drop.html.

31. "Clarks Rule and Youngs Rule, Calculating Pediatric and Adult Dosages," Web, http://www.pharmacy-tech-study.com/dosecalculation.html.

32. R. Nave, Web, http://hyperphysics.phy-astr.gsu.edu/hbase/sound/radar.html.

33. "American Experience: TV's Most-watched History Series," PBS, Web, http://www.pbs.org/wgbh/americanexperience/features/primary-resources/goldengate-true-picture/.

34. James Lyons, "Crypto," Practical Graphy, Web, http://practicalcryptography.com/ciphers/hill-cipher/.

Chapter IV

1. "What Speed Should I Drive to Get Maximum Fuel Efficiency?" 27 September 2000, Web, HowStuffWorks.com, http://auto.

howstuffworks.com/fuel-efficiency/fuel-economy/question477.htm.

2. "How Does the Calculator Find Values of Sine (or Cosine or Tangent)?" Web, http://www.homeschoolmath.net/teaching/sine_calculator.php.

3. The Editors of Encyclopædia Britannica. "Henry Briggs," Encyclopedia Britannica Online, Web, http://www.britannica.com/biography/Henry-Briggs.

4. "Interest Rate Formulas," 10 Dec. 1998, Web, https://www.math.nmsu.edu/~pmorandi/math210gs99/InterestRateFormulas.html.

5. "Hanford Isotope Project Strategic Business Analysis Cesium-137 (Cs-137)," 18 Aug. 2005, Web, http://www.bt.cdc.gov/radiation/isotopes/pdf/cesium.pdf.

6. "Apparent Magnitude—COSMOS," Web, http://astronomy.swin.edu.au/cosmos/A/Apparent+Magnitude.

7. "Definition of IQ," Web, http://hiqnews.megafoundation.org/Definition_of_IQ.html.

8. "Group Profile," *Bee World* 43.4 (1962): 131–33, Web, https://secure-media.college-board.org/digitalServices/pdf/sat/total-group-2015.pdf.

9. Michael Lemmon, "What Is an RC Circuit?" 1 Feb. 2009, Web, http://www3.nd.edu/~lemmon/courses/ee224/web-manual/web-manual/lab8a/node4.html.

10. Marshall Brain, "How Carbon-14 Dating Works," 3 October 2000, Web, HowStuffWorks.com, http://science.howstuffworks.com/environmental/earth/geology/carbon-14.htm.

11. "The Catenary—National Curve Bank," Web, http://curvebank.calstatela.edu/catenary/catenary.htm.

12. "Ideal Rocket Equation," 12 June 2014, Web, https://spaceflightsystems.grc.nasa.gov/education/rocket/rktpow.html.

13. Stephen Hawking, *A Brief History of Time* (New York: Bantam Dell, 1988), page 139.

14. "Introduction to Fractal Geometry," Web, http://www.fractal.org/Bewustzijns-Besturings-Model/Fractals-Useful-Beauty.htm.

15. "Fractal Image Compression," Web, http://www.math.psu.edu/tseng/class/Fractals.html.

16. Murray Bourne, "Polar Coordinates and Cardioid Microphones," Intmath.com RSS, Web, http://www.intmath.com/blog/mathematics/polar-coordinates-and-cardioid-microphones-2496.

17. Murray Bourne, "Golden Spiral," Intmath.com RSS, Web, http://www.intmath.com/blog/mathematics/golden-spiral-6512.

18. "How Does the Calculator Find Values of Sine (or Cosine or Tangent)?" Web, http://www.homeschoolmath.net/teaching/sine_calculator.php.

19. "All About Aspheric Lenses," Web, http://www.edmundoptics.com/resources/application-notes/optics/all-about-aspheric-lenses/.

20. Clark Kimberling, "1972 Meeting with AAAS in Washington, D.C," *Bulletin of the Ecological Society of America* 53.4 (1972): 6–29, Web, http://faculty.evansville.edu/ck6/ellipse.pdf.

21. Natasha Glydon, "Lithotripsy: A Medical Application of the Ellipse," Math Central, Web, http://mathcentral.uregina.ca/beyond/articles/Lithotripsy/lithotripsy1.html.

22. Chris Baker, "Elliptical Pool: An Even Nerdier Sport Than Quidditch," Wired.com, Conde Nast Digital, 27 Nov. 2015, Web, http://www.wired.com/2015/11/elliptic-pool-loop-round-billiard-table/.

23. Dan Daley, "Networks' Audio Sounds Some New Notes for NFL Season," Sports Video Group, 13 Sept. 2012, Web, http://www.sportsvideo.org/2012/09/13/networks-audio-sounds-some-new-notes-for-nfl-season/.

24. Golden Gate Bridge Data, Web, http://goldengatebridge.org/research/factsGGBDesign.php.

25. Elizabeth Howell, "What Is the Vomit Comet?" LiveScience, TechMedia Network, 30 Apr. 2013, Web, http://www.livescience.com/29182-what-is-the-vomit-comet.html.

26. "Cooling Towers," 20 Feb. 2006, Web, http://www.nucleartourist.com/systems/ct.htm.

27. "The Mathematics of Loran," Math Central, Web, http://mathcentral.uregina.ca/beyond/articles/LoranGPS/Loran.html.

28. LORAN Legislation, 7 Mar. 2016, Web, http://www.gps.gov/policy/legislation/loran-c/.

29. Joseph A. Shaw, "Reflector Telescopes Reflector Telescope Design," Web, http://www.montana.edu/jshaw/documents/17a%20EELE582_S15_TelescopeDesign.pdf

30. "Trajectories and Orbits," Web, http://history.nasa.gov/conghand/traject.htm.

Chapter V

1. "Home Entertainment," THX.com HDTV Set Up Comments, Web, http://www.thx.com/consumer/home-entertainment/home-theater/hdtv-set-up/.

2. Tim Plaehn, "Approach Angle and Aerodynamics: How Plane Aerodynamics Work," Bright Hub, 18 Aug. 2010, Web, http://www.brighthub.com/science/aviation/articles/82974.aspx.

3. Sarah Pruitt, "Why Does the Leaning Tower of Pisa Lean?" History.com, A&E Television Networks, 03 Dec. 2015, Web, http://www.history.com/news/ask-history/why-does-the-leaning-tower-of-pisa-lean.

4. The Great Trigonometric Survey, Web, http://livelystories.com/2012/01/23/the-great-trigonometric-survey/.

5. "Six Sundial Projects for You to Make," Sundials on the Internet, 21 Dec. 2013, Web, http://www.sundials.co.uk/projects.htm.

6. "Luminous Intensity," SpringerReference, Web, http://www.sharp-world.com/contents/calculator/support/program/pdf/el9900/Cal_06_EL9900.pdf.

7. "Distance: Direction Calculating Between Two Location on the Planet Directly on the Map," Web, http://www.sunearthtools.com/tools/distance.php#txtDist_1.

8. "Cosine Effect Error," Police Radar, Web, http://copradar.com/preview/chapt2/ch2d1.html.

9. "Athletics," Web, http://www.physicsclassroom.com/class/circles/Lesson-2/Athletics.

10. Eric W. Weisstein, "Mercator Projection," From MathWorld—A Wolfram Web Resource, Web, http://mathworld.wolfram.com/MercatorProjection.html.

11. "How Do Active Noise-cancelling Headphones Work?" Audio-Technica US, Web, http://www.audio-technica.com/cms/features/b3ef06fca462fcad/.

12. "What Do Mufflers Do? Learn How Mufflers Work," AutoAccessoriesGarage.com, Web, http://www.autoaccessoriesgarage.com/Exhaust/What-Do-Mufflers-Do.

13. "Fundamental Frequency and Harmonics," Web, http://www.physicsclassroom.com/class/sound/Lesson-4/Fundamental-Frequency-and-Harmonics.

14. Matt Weschler, "How Lasers Work," 1 April 2000, Web, HowStuffWorks.com, http://science.howstuffworks.com/laser.htm.

15. Ron Kurtus, "Background of Worldwide AC Voltages and Frequencies," Web, http://www.school-for-champions.com/science/ac_world_volt_freq.htm#.VxZL3fkrI7Z.

Bibliography

ADA Ramp Codes. http://www.guldmann.net/Files/Billeder/GuldmannProdukter/ Stepless/Transportable%20ramper/Shared%20files/US/ADA%20Ramp%20Codes_ US.pdf.

"Adjusting Football's Pythagorean Theorem." Pro-football-reference.com. Blog. 18 July 2007. Web. http://www.pro-football-reference.com/blog/?p=337.

"All About Aspheric Lenses." Web. http://www.edmundoptics.com/resources/applica tion-notes/optics/all-about-aspheric-lenses/.

"*American Experience*: TV's Most-watched History Series." PBS. Web. http://www.pbs. org/wgbh/americanexperience/features/primary-resources/goldengate-truepic ture/.

"Apparent Magnitude | COSMOS." Web. http://astronomy.swin.edu.au/cosmos/A/Ap parent+Magnitude.

"Astronomy 101 Specials: Aristarchus and the Size of the Moon." Web. http://www.eg. bucknell.edu/physics/astronomy/astr101/specials/aristarchus.html.

"Athletics." Web. http://www.physicsclassroom.com/class/circles/Lesson-2/Athletics.

Baker, Chris. "Elliptical Pool: An Even Nerdier Sport Than Quidditch." Wired.com. Conde Nast Digital. 27 Nov. 2015. Web. http://www.wired.com/2015/11/elliptic-pool-loop-round-billiard-table/.

Barrows, John D. *100 Essential Things You Didn't Know You Didn't Know About Math & the Arts*. New York: W.W. Norton, 2014.

Bellos, Alex. *Here's Looking at Euclid*. New York: Simon & Schuster, 2010.

Bellos, Alex. *The Grapes of Math*. New York: Simon & Schuster, 2014.

Benson, Tom, "Lift Equation—Activity." 12 June 2014. Web. http://www.grc.nasa.gov/ WWW/k-12/BGA/Melissa/lift_equation_act.htm.

"BMI Calculator." Body Mass Index. Web. http://bmi-calories.com/bmi-calculator.html.

Bourne, Murray. "Golden Spiral." Intmathcom RSS. Web. http://www.intmath.com/blog/ mathematics/golden-spiral-6512.

Bourne, Murray. "Polar Coordinates and Cardioid Microphones." Intmathcom RSS. Web. http://www.intmath.com/blog/mathematics/polar-coordinates-and-cardioid-microphones-2496.

Boyer, Carl B. *A History of Mathematics*. New York: John Wiley & Sons, 1968.

Brain, Marshall. "How Bicycles Work." 1 April 2000. Web. HowStuffWorks.com. http:// adventure.howstuffworks.com/outdoor-activities/biking/bicycle.htm.

Brain, Marshall. "How Carbon-14 Dating Works." 3 October 2000. Web. HowStuffWorks. com. http://science.howstuffworks.com/environmental/earth/geology/carbon-14.htm.

Brain, Marshall, and Tom Harris. "How GPS Receivers Work." 25 September 2006. Web.

HowStuffWorks.com. http://electronics.howstuffworks.com/gadgets/travel/gps. htm.

Cars Traveling Around a Banked Curve. Web. http://plaza.obu.edu/corneliusk/up1/ bc_f.pdf.

"The Catenary—National Curve Bank." Web. http://curvebank.calstatela.edu/catenary/ catenary.htm.

"Chapter 6 Fats." 9 July 2008. Web. http://health.gov/dietaryguidelines/dga2005/docu ment/html/chapter6.htm.

"Clarks Rule and Youngs Rule, Calculating Pediatric and Adult Dosages." Web. http:// www.pharmacy-tech-study.com/dosecalculation.html.

Comments. TVline.com. 29 May 2014. Web. http://tvline.com/2014/05/29/tv-ratings-nielsen-to-increase-sample-size/.

"Cooling Towers." Cooling Towers. 20 Feb. 2006. Web. http://www.nucleartourist.com/ systems/ct.htm.

"Cosine Effect Error." Police Radar. Web. http://copradar.com/preview/chapt2/ch2d1.html.

Daley, Dan. "Networks' Audio Sounds Some New Notes for NFL Season." Sports Video Group. 13 Sept. 2012. Web. http://www.sportsvideo.org/2012/09/13/networks-audio-sounds-some-new-notes-for-nfl-season/.

Davenport, Clay, and Keith Woolner. "Baseball Prospectus—Revisiting the Pythagorean Theorem." Baseball Prospectus. 30 June 1999. Web. http://www.baseballprospectus. com/article.php?articleid=342.

Davis, Doug. "Circular Motion (and Other Things)." 2001. Web. http://www.ux1.eiu. edu/~cfadd/1350/06CirMtn/FlatCurve.html.

"Definition of IQ." Web. http://hiqnews.megafoundation.org/Definition_of_IQ.html.

"Design and Construction of Aquaculture Facilities." Aquaculture Engineering: 294–320. Web. http://www.phy6.org/stargaze/Sarist.htm.

"Distance." Direction Calculating Between Two Location on the Planet Directly on the Map. Web. http://www.sunearthtools.com/tools/distance.php#txtDist_1.

Durgun, M. Erk. "6. Distance—Time—Relativity." Space-Time Interval. Web. http:// www.unitytheory.info/space-time_interval.html.

"Earth Atmosphere Model—Imperial Units." Ed. Nancy Hall. Web. http://www.grc.nasa. gov/WWW/k-12/airplane/atmos.html.

The Editors of Encyclopædia Britannica. "Henry Briggs." Encyclopedia Britannica Online. Web. http://www.britannica.com/biography/Henry-Briggs.

"Employment Situation News Release." U.S. Bureau of Labor Statistics. Web. http:// www.bls.gov/news.release/empsit.htm.

"Eratosthenes." Famous Scientists. famousscientists.org. 21 Jun. 2014. Web. http://www. famousscientists.org/eratosthenes/.

"The Fifth Postulate." Web. http://www.physicscentral.com/explore/writers/bardi.cfm.

"Formula Used to Calculate Wind Chill." USATODAY.com. 30 Oct. 2001. Web. http:// usatoday30.usatoday.com/weather/winter/windchill/wind-chill-formulas.htm.

"Fractal Image Compression." Web. http://www.math.psu.edu/tseng/class/Fractals.html.

Freiberger, Marianne. "Maths Behind the Rainbow." Plus.maths.org. Web. https://plus. maths.org/content/rainbows.

"Fundamental Frequency and Harmonics." Web. http://www.physicsclassroom.com/ class/sound/Lesson-4/Fundamental-Frequency-and-Harmonics.

"Gas Laws." Web. http://chemistry.bd.psu.edu/jircitano/gases.html.

Gibilisco, Stan. Physics Demystified. New York: McGraw-Hill, 2007.

Gibilisco, Stan. Teaching Yourself Electricity and Electronics. 5th ed. New York: McGraw-Hill, 2011.

Glydon, Natasha, "Lithotripsy—A Medical Application of the Ellipse—Math Central." Web. http://mathcentral.uregina.ca/beyond/articles/Lithotripsy/lithotripsy1.html.

Godbole, Medha. "Official Basketball Size." Buzzle.com. 5 June 2015. Web. http://www.buzzle.com/articles/official-basketball-size.html.

Golden Gate Bridge Data. Web. http://goldengatebridge.org/research/factsGGBDesign.php.

Gray, Todd. "Tolkowsky Ideal Cut Diamond Proportions & Tolkowsky Diamond Design." Nice Ice Diamonds Comments. 2015. Web. http://niceice.com/tolkowsky-ideal-cut/.

The Great Trigonometric Survey. Web. http://livelystories.com/2012/01/23/the-great-trigonometric-survey/.

"Group Profile." *Bee World* 43.4 (1962): 131–33. Web. https://secure-media.collegeboard.org/digitalServices/pdf/sat/total-group-2015.pdf.

Hachman, Mark. "Nvidia Finally Gets Faces Right—Until They Open Their Mouths—ReadWrite." ReadWrite. 21 Mar. 2013. Web. http://readwrite.com/2013/03/21/nvidia-finally-gets-faces-right-until-they-open-their-mouths/.

Hadley, Debbie. "How to Use Crickets to Calculate Temperature." About.com Education. Web. http://insects.about.com/od/insectfolklore/a/crickets-temperature.htm.

"Hanford Isotope Project Strategic Business Analysis Cesium-137 (Cs-137)." 18 Aug. 2005. Web. http://www.bt.cdc.gov/radiation/isotopes/pdf/cesium.pdf.

Harris, Tom. "How Cameras Work." 21 March 2001. Web. HowStuffWorks.com. http://electronics.howstuffworks.com/camera.htm.

Hartley, Simon, "The BMI Formula." How to Calculate. Web. http://www.whathealth.com/bmi/formula.html.

Harvard School of Engineering and Applied Sciences—Cs 152: Programming Language. Curry-Howard Isomorphism. 2 Apr. 2013. Web. http://www.seas.harvard.edu/courses/cs152/2013sp/lectures/lec17.pdf.

Hawking, Stephen. *A Brief History of Time.* New York: Bantam Dell, 1988.

Helmenstein, Anne Marie. "Learn the Chemistry Behind Gasoline and Octane Ratings." About.com Education. 10 Oct. 2015. Web. http://chemistry.about.com/cs/howthingswork/a/aa070401a.htm.

Helmenstein, Anne Marie. "What Are the Types of Crystals?" About.com Education. 2 Dec. 2014. Web. http://chemistry.about.com/cs/growingcrystals/a/aa011104a.htm.

Herring, Jon. "Sold to the Highest Bidder—the Fatally Flawed Food Pyramid." Early to Rise. Web. http://www.earlytorise.com/our-no-hassle-plan-for-healthy-living/sold-to-the-highest-bidder-the-fatally-flawed-food-pyramid/.

"Home Entertainment." THX.com. HDTV Set Up Comments. Web. http://www.thx.com/consumer/home-entertainment/home-theater/hdtv-set-up/.

Horton, Sarah, and Patrick Lynch, "Color Displays." GRAPHICS. 2002. Web. http://webstyleguide.com/wsg2/graphics/displays.html.

"How Do Active Noise-cancelling Headphones Work?" Audio-Technica US. Web. http://www.audio-technica.com/cms/features/b3ef06fca462fcad/.

"How Do I Locate That Earthquake's Epicenter?" Web. http://www.geo.mtu.edu/UPSeis/locating.html.

"How Does the Calculator Find Values of Sine (or Cosine or Tangent)?" Web. http://www.homeschoolmath.net/teaching/sine_calculator.php.

"How the Government Measures Unemployment." U.S. Bureau of Labor Statistics. U.S. Bureau of Labor Statistics. 8 Oct. 2015. Web. http://www.bls.gov/cps/cps_htgm.htm.

"How to Calculate Boiling Point—Video & Lesson Transcript." Study.com. Web. http://study.com/academy/lesson/how-to-calculate-boiling-point.html#transcriptHeader.

"How to Calculate the Energy Available from Foods." TimetoRun Nutrition. Web. http://www.time-to-run.com/nutrition/calculate-energy.htm.

Howell, Elizabeth. "What Is the Vomit Comet?" LiveScience. TechMedia Network. 30 Apr. 2013. Web. http://www.livescience.com/29182-what-is-the-vomit-comet.html.

"Ideal Rocket Equation." 12 June 2014. Web. https://spaceflightsystems.grc.nasa.gov/education/rocket/rktpow.html.

Impey, Chris. "Nuclear Reactions in the Sun." 2012. Web. http://m.teachastronomy.com/astropedia/article/Nuclear-Reactions-in-the-Sun#.

"Interest Rate Formulas." 10 Dec. 1998. Web. https://www.math.nmsu.edu/~pmorandi/math210gs99/InterestRateFormulas.html.

"Introduction to Fractal Geometry." Web. http://www.fractal.org/Bewustzijns-Besturings-Model/Fractals-Useful-Beauty.htm.

"Is Your Roof Angle Suitable for Solar Panels?" Which Solar Panel Angle Is Best? Web. https://www.solarquotes.com.au/panels/angle/.

Kaiser, Gary E. "Size and Shapes of Viruses." June 2015. Web. http://faculty.ccbcmd.edu/courses/bio141/lecguide/unit4/viruses/ssvir.html.

"Kepler's Three Laws." Web. http://www.physicsclassroom.com/class/circles/Lesson-4/Kepler-s-Three-Laws.

"Kilobytes Megabytes Gigabytes." Web. https://web.stanford.edu/class/cs101/bits-giga bytes.html.

Kimberling, Clark. "1972 Meeting with AAAS in Washington, D.C." *Bulletin of the Ecological Society of America* 53.4 (1972): 6–29. Web. http://faculty.evansville.edu/ck6/ellipse.pdf.

Kline, Morris. *Mathematics for the Nonmathematician.* New York: Dover Publications, 1967.

Kroto, Harold W. "Fullerene." Encyclopedia Britannica Online. 11 Nov. 2014. Web. http://www.britannica.com/science/fullerene.

Kurtus, Ron. "Background of Worldwide AC Voltages and Frequencies." Web. http://www.school-for-champions.com/science/ac_world_volt_freq.htm#.VxZL3fkrI7Z.

Lantigua-Williams, Juleyka. "Are You in the Shrinking Middle Class?" *The Atlantic.* 9 Dec. 2015. Web. http://www.theatlantic.com/politics/archive/2015/12/are-you-in-the-shrinking-middle-class-take-this-2-step-test/433599/.

"Laying Out Applications." Pro PHP-GTK: 87–118. Web. http://www.pntf.org/officials/ivars/Throwing_Event_Sector_Angles_Rev_F1.pdf.

Lemmon, Michael. "What Is an RC Circuit?" 1 Feb. 2009. Web. http://www3.nd.edu/~lemmon/courses/ee224/web-manual/web-manual/lab8a/node4.html.

"Lens Maker's Formula." Tutorvista.com. Web. http://www.tutorvista.com/content/physics/physics-iv/optics/lens-makers-formula.php.

LORAN Legislation. 7 Mar. 2016. Web. http://www.gps.gov/policy/legislation/loran-c/.

Louie, J. "What Is Richter Magnitude?" 9 Oct. 1996. Web. http://crack.seismo.unr.edu/ftp/pub/louie/class/100/magnitude.html.

"Luminous Intensity." SpringerReference. Web. http://www.sharp-world.com/contents/calculator/support/program/pdf/el9900/Cal_06_EL9900.pdf.

Lyons, James. "Crypto." Practical Graphy. Web. http://practicalcryptography.com/ciphers/hill-cipher/.

Madison, N., and Bronwyn Harris. WiseGeek. Conjecture. 7 Apr. 2016. Web. http://www.wisegeek.com/what-is-supply-and-demand.htm.

"The Mathematics of Loran." Math Central. Web. http://mathcentral.uregina.ca/beyond/articles/LoranGPS/Loran.html.

Maxwell, Alison. "Supreme Court Rejects Census Sampling Plan." *Government Executive.* 25 Jan. 1999. Web. http://www.govexec.com/federal-news/1999/01/supreme-court-rejects-census-sampling-plan/1597/.

McAuley, David. "Mean Arterial Pressure—MAP Calculator." Web. http://www.globalrph.com/map.cgi.

Mitovich, Matt Webb. "Number of 'Nielsen Families' That Fuel TV Ratings to Grow 'Significantly.'" TVLine.com.

Nave, R. Web. http://hyperphysics.phy-astr.gsu.edu/hbase/sound/radar.html.

"NCAA and NFL Passing Efficiency Computation." NCAA and NFL Passing Efficiency Computation. Web. http://football.stassen.com/pass-eff/.

NFL Rulebook. Web. http://operations.nfl.com/the-rules/2015-nfl-rulebook/#rule-2.-the-ball.

Olivier, Alwyn. "2C." Feb. 1999. Web. http://academic.sun.ac.za/mathed/174/circlesregionschords.pdf.

"OpenAnesthesia." Web. https://www.openanesthesia.org/poiseuilles_law_iv_fluids/.

Pickover, Clifford A. *Archimedes to Hawking*. Oxford: Oxford University Press, 2008.

Plaehn, Tim. "Approach Angle and Aerodynamics—How Plane Aerodynamics Work." Bright Hub. 18 Aug. 2010. Web. http://www.brighthub.com/science/aviation/articles/82974.aspx.

Pruitt, Sarah, "Why Does the Leaning Tower of Pisa Lean?" History.com. A&E Television Networks. 03 Dec. 2015. Web. http://www.history.com/news/ask-history/why-does-the-leaning-tower-of-pisa-lean.

"Pythagorean Expectation Calculator & Formula." Pythagorean Expectation Calculator. Web. http://www.had2know.com/sports/pythagorean-expectation-win-percentage-baseball.html.

"QB Rating Calculator." Web. http://brucey.net/nflab/statistics/qb_rating.html.

Roos, Dave. "How Computer Animation Works." 28 April 2008. Web. HowStuffWorks.com. http://entertainment.howstuffworks.com/computer-animation.htm.

Rouse, Margaret. "What Is X and Y Coordinates? Definition from WhatIs.com." WhatIs.com. Apr. 2005. Web. http://whatis.techtarget.com/definition/x-and-y-coordinates.

"Saffir-Simpson Hurricane Wind Scale." Web. http://www.nhc.noaa.gov/aboutsshws.php.

Schmidt, Timothy, and David Baxter. "The Physics of Scuba Diving." 2003. Web. http://physics.itsbaxter.com/pressure.html.

Schmitt, Barton D. "Acetaminophen (Tylenol) Dosage Table." Kid Care: Acetaminophen Dosage Chart for Children and Infants. Web. http://www.stlouischildrens.org/articles/kidcare/acetaminophen-tylenol-etc-dosage-table.

"A Selection from Symbolic Logic. Lewis Carroll." YesFine.com. Web. http://www.yesfine.com/carroll_symbolic_logic.htm.

Serkis, Andy. "Gollum in The Lord of the Rings—Performance Capture." 2016. Web. http://www.serkis.com/performance-capture-gollum.htm.

Shaw, Joseph A. "Reflector Telescopes Reflector Telescope Design." Web. http://www.montana.edu/jshaw/documents/17a%20EELE582_S15_TelescopeDesign.pdf.

Simmons, John Galbraith. *The Scientific 100*. New York: Fall River Press, 1996.

"Single-Season Leaders & Records for Power-Speed." Baseball-Reference.com. Web. http://www.baseball-reference.com/leaders/power_speed_number_season.shtml.

"Six Sundial Projects for You to Make." Sundials on the Internet. 21 Dec. 2013. Web. http://www.sundials.co.uk/projects.htm.

Skorseth, Ken, and Ali A. Selim. "Gravel Roads: Maintenance and Design Manual." Washington, D.C.: U.S. Dept. of Transportation, Federal Highway Administration. 2000. Web. https://www.epa.gov/sites/production/files/2015–10/documents/2003_07_24_nps_gravelroads_gravelroads.pdf.

Smith, Martha K. "History of Negative Numbers." M 326K. 19 Feb. 2001. Web. https://www.ma.utexas.edu/users/mks/326K/Negnos.html.

Spathopoulos, Vassilios. "Coefficient of Restitution." In Sports. Web. http://www.topendsports.com/biomechanics/coefficient-of-restitution.htm.

Stern, David P. "(8) The Round Earth and Christopher Columbus." 31 Mar. 2014. Web. http://www.phy6.org/stargaze/Scolumb.htm.

Stern, David P. "(9a) May Earth Be Revolving Around the Sun?" 2 Apr. 2014. Web. http://www.phy6.org/stargaze/Sarist.htm.

"Stopping Distance Formula." Web. http://www.softschools.com/formulas/physics/stopping_distance_formula/89/.

"STR: A Brief History of Einstein's Special Theory of Relativity." Web. http://www.twow.net/ObjText/OtkCaLbStrB.htm.

Suits, B.H., "Pythagorean Scale." Web. http://www.phy.mtu.edu/~suits/pythagorean.html.

"10 Amazing Facts About the Geodesic Sphere of Spaceship Earth." DisneyFanatic.com. 11 Sept. 2014. Web. http://www.disneyfanatic.com/10-amazing-facts-geodesic-sphere-spaceship-earth/.

"Throw Ratios and Viewing Distances." Web. http://www.theprojectorpros.com/learn-s-learn-p-theater_throw_ratios.htm.

"Tolerances for Rectangular Bars." Tolerance Values. Web. http://www.alumeco.com/Knowledge-and-Technique/Tolerances/Tolerances-for-rectangular-bars.aspx.

Toshiba Website. "Television's Knowledge Base. How To: Best Distance and Viewing Angle." 24 Apr. 2013. Web. http://forums.toshiba.com/t5/Televisions-Knowledge-Base/How-To-Best-distance-and-viewing-angle-for-your-HDTV/ta-p/281923.

"Trajectories and Orbits." Web. http://history.nasa.gov/conghand/traject.htm.

"Tuning and Ratio." Nrich.maths.org. Web. https://nrich.maths.org/5453.

United States National Park Service. "Gateway Arch Fact Sheet." National Parks Service. U.S. Department of the Interior. Web. https://www.nps.gov/jeff/planyourvisit/gateway-arch-fact-sheet.htm.

"Volume/Time—IV Drop Rate Questions." DosageHelp.com. Web. http://www.dosagehelp.com/iv_rate_drop.html.

Waehner, Paige, "Understanding Your Maximum Heart Rate." Verywell. 14 Jan. 2016. Web. http://exercise.about.com/od/healthinjuries/g/maxheartrate.htm.

Wall, Mike. "The Most Extreme Human Spaceflight Records." Space.com. Web. http://www.space.com/11337-human-spaceflight-records-50th-anniversary.html.

Weisstein, Eric W. "Mercator Projection." From *MathWorld*—A Wolfram Web Resource. Web. http://mathworld.wolfram.com/MercatorProjection.html.

"Welcome to VMAR (Vehicle Mechanical and Accident Reconstruction, LLC)!" June 2011. Web. http://www.vmar.net/email_jun11.htm.

Weschler, Matt, "How Lasers Work." 1 April 2000. Web. HowStuffWorks.com. http://science.howstuffworks.com/laser.htm.

"What Do Mufflers Do? Learn How Mufflers Work." AutoAccessoriesGarage.com. Web. http://www.autoaccessoriesgarage.com/Exhaust/What-Do-Mufflers-Do.

"What Is Fowler's Position? (Definition and Explanation)." www.nursefrontier.com. 25 Jan. 2016. Web. http://www.nursefrontier.com/what-is-fowlers-position-definition-and-explanation/.

"What Is Seismology and What Are Seismic Waves?" Web. http://www.geo.mtu.edu/UPSeis/waves.html.

"What Speed Should I Drive to Get Maximum Fuel Efficiency?" 27 September 2000. Web. HowStuffWorks.com. http://auto.howstuffworks.com/fuel-efficiency/fuel-economy/question477.htm.

Zebrowski, Jr., Ernest. *A History of the Circle.* New Brunswick, NJ: Rutgers University Press, 2000.

Zimmermann, Kim Ann. "Temperature: Facts, History & Definition." LiveScience. Tech-Media Network. 20 Sept. 2013. Web. http://www.livescience.com/39841-temperature.html.

INDEX